KV-308-857

GEOGRAPHY

An Outline for the Intending Student

CONTRIBUTORS

W. G. V. Balchin

Alice Coleman

W. Gordon East

K. C. Edwards

S. Gregory

John W. House

Cuchlaine A. M. King

W. R. Mead

J. Oliver

Robert W. Steel

GEOGRAPHY

An Outline for the Intending Student

Edited by

Professor W. G. V. Balchin

Head of the Department of Geography,
University College of Swansea

LONDON
ROUTLEDGE & KEGAN PAUL

First published 1970
by Routledge & Kegan Paul Ltd
Broadway House, 68–74 Carter Lane
London, E.C.4
Printed in Great Britain
by Willmer Brothers Limited
Birkenhead, Cheshire

ISBN 0 7100 6824 7 (C)
ISBN 0 7100 6825 5 (P)

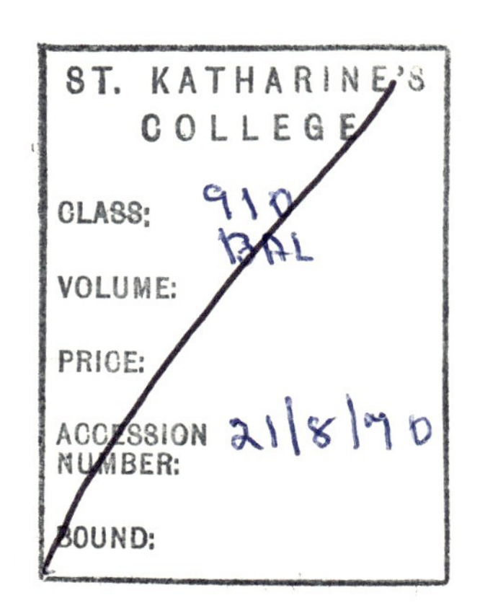

CONTENTS

Preface vii

1. Introduction: The Nature and Content of Geography 1
W. G. V. BALCHIN

The Basic Skills of Geography

2. Literacy 12
W. R. MEAD
Professor of Geography, University College, London

3. Graphicacy 28
W. G. V. BALCHIN

4. Numeracy 43
S. GREGORY
Professor of Geography, University of Sheffield

The Organization of Geographical Material

5. Field Work 55
ALICE COLEMAN
Reader in Geography, King's College, London

6. Hypotheses and Models 71
CUCHLAINE A. M. KING
Professor of Geography, University of Nottingham

7. Systematic Geography 86

W. GORDON EAST
Professor of Geography,
Birkbeck College, London

8. Regional Geography 99

K. C. EDWARDS, C.B.E
Professor of Geography,
University of Nottingham

Pure and Applied Geography

9. Land Use and Resource Analysis 115

ALICE COLEMAN

10. Applied Geography 132

JOHN W. HOUSE
Professor of Geography,
University of Newcastle-upon-Tyne

Information for the Intending Geographer

11. Geography in British Universities 147

ROBERT W. STEEL
Professor of Geography,
University of Liverpool

12. Careers in Geography 163

J. OLIVER
Professor of Geography,
University College, Swansea

APPENDIX: Additional Information 181

W. G. V. BALCHIN

Index 193

PREFACE

During the last decade there has been a significant advance in the scope and orientation of university geography. New methods of collecting and analysing data, greater precision afforded by quantitative techniques, the development of computer-graphics, new conceptual frameworks of thought, and the acceptance of a geographical approach to the solution of social and economic problems—all these, and more, mean that geography is entering the 1970s with expanding horizons undreamt of at mid-century.

This book outlines the basic nature and content of modern geography and describes the transformation that has taken place in the subject. It aims to provide an integrated account of how geography is now studied in the universities of the United Kingdom.

The primary object of the book is to provide guidance for the potential geography student. It should also enable the committed student to fit the component parts of geography more readily into an overall context. It is also hoped that the book will be of assistance to geography teachers anxious to keep themselves abreast of developments in their subject.

W. G. V. BALCHIN

I

Introduction: The Nature and Content of Geography

W. G. V. Balchin

Geography as a university subject has been steadily attracting more students in the post-war period. During the last decade there have sometimes been more sixth-formers applying to study geography than any other subject, and in every year it has been well to the forefront of demand by intending students. It is clearly a major, although little publicized, academic magnet, and for this there are many good reasons.

In the first place, geography is concerned with the reality of the human habitat, the planet earth that we all have to live on and find a living on. This is the century of a shrinking globe and an exploding population, when competition for the planet's space and resources have become fiercer than ever before and of immediate concern to everyone. In 1969, the first photographs of the whole earth as a single unit, taken from the Apollo 10 lunar flight, became available and these have brought home, even to the man in the street, what the great British geographer Mackinder was saying as early as 1924: 'the habitat of each separate human being is this global earth.' It is increasingly difficult to escape from the effects of major events or geographical changes in apparently remote parts of the earth; repercussions far afield from the source may be both prompt and radical throughout the world. Consequently, the study of geography becomes increasingly vital and relevant in the modern age.

Second, geography is an extremely varied and versatile subject. Because its unifying theme is the significance of place and of spatial relationships, it can touch upon any topic that contributes to the nature of places, analysing their distribution patterns, the causes of these patterns and the effects that they produce. These topic studies, collectively referred to as Systematic Geography, give great breadth and variety to the subject, so that students are attracted into it at many different points of entry.

Third, this variety means that geography shares common frontiers with many other subjects. For example, botany studies plant life and economics studies the production, exchange and consumption of scarce commodities. Geography is interested in both these topics from the viewpoint of distribution over the face of the earth, and their differentiation from place to place; it studies vegetation and economic activity as two of the many branches of systematic geography. As a result of its common interests with other subjects, geography is strategically placed to participate in one of the most active academic phenomena of our times—the growth of new knowledge at the frontiers between related mental disciplines. In each case, while geography has been enriched by contact with the cognate subject's body of theory, it has also been able to illuminate the cognate subject by its own distinctive spatial approach. It is no accident that a comprehensive field survey of Britain's vegetation should have been made by geographers rather than botanists, that geographers rather than doctors have mapped the distribution of diseases, that geographers rather than economists should have urged the establishment of the Water Resources Board, nor that geographers have an interest in air pollution studies. Geographers had been studying water and air long before these became scarce commodities of economic significance. These examples illustrate the great scope for new discovery in geography. It is essentially a developing subject.

Fourth, as well as the systematic branches that share common interests with other disciplines, geography also contains Regional Studies which draw all the developing ideas inward and integrate them in relation to specific places.

Regional geography is concerned with the explanation, even more fully and satisfying, of why places should be so distinctive and why man's uses of them should be so different. Regional geography is to systematic geography as the brain is to the peripheral sensory nerves. It absorbs their discoveries and co-ordinates them. New knowledge and new theory are dovetailed into place and applied with selective relevance to regions at all scales, and even to the whole world as a unit. This integrative role of regional geography tends to have an integrative effect upon its students. Not for them is the narrow path of specialism. Rather they aim to become well-rounded mentally, with minds alert and open to new facets of understanding that may at any time modify the existing complex of truth.

Fifth, in the last ten or twelve years geography has been developing greater precision through improved quantification. Like the social sciences, it has never lent itself to exact measurement in the manner of the physical sciences, and until recently this has been something of a handicap. Light always travels at a constant speed, but there are no phenomena in geography that behave as consistently. This is because geographical phenomena are more complex, the result of many interacting factors; it is also less easy to isolate any given factor because we do not necessarily know all the other factors from which it must be isolated. However, this is just the sort of situation that lends itself to probability analysis leading on to statements as to how great a part of the distribution is to be explained by factors not yet identified. Geography is making great strides in applying such techniques, and today's student has the pleasure of seeing his subject become progressively more precise even as he studies it.

Finally, because of its advancement in both theoretical understanding and quantitative precision, geography has become a powerful tool for analysing and solving problems. Its practical value is being increasingly recognized by governments at all levels, and geographers are being called into consultation more frequently than ever before. At local government level in this country geographers are proving their value in the planning field, while at the other end of the scale the United Nations Organization is commissioning more geographers to tackle the problems of underdeveloped countries. The growth of employment opportunities for geographers has outstripped even the growth in the number of geographers available. The demand is high and the opportunities are varied.

Geography, therefore, is a subject which can be of great value in the personal mental development of the student and also in the development of the planet that he studies. It offers almost unlimited scope, but what it does not do is to guarantee the realization of its benefits without effort on the part of the student. It demands active work and hard constructive thinking before the student can effectively pass through the gateway that it opens.

Geography is concerned with places, environments and distributions. It seeks the relationships between classes of data, discusses systems of organization, and probes transformations and implications. Its sphere is the whole world. Because this involves such a vast sum of knowledge, divisions of the subject are inevitable, and these are organized within the two major aspects already mentioned: systematic and regional.

Systematic geography involves studying the distribution of single phenomena or groups of phenomena over the surface of the earth. In earlier times the emphasis was on natural phenomena, now constituting Physical Geography. More recently cultural phenomena have also become important, and these constitute Human Geography. Both these basic divisions, physical and human, are again subdivided into recognized fields of knowledge.

Physical Geography includes the study of the four spheres: the atmosphere, which is the subject of Climatology and Meteorology; the hydrosphere, the subject of Hydrology and Oceanography; the lithosphere, covered by Geomorphology and the biosphere covered by Biogeography. The last of these is again divided into three aspects, Plant Geography, Animal Geography and Pedology, the study of soils.

In Human Geography the subdivision of aspects is not yet generally agreed, and varies from university to university. It seems to be moving towards a basic two-fold grouping of Social Geography on the one hand and Economic Geography on the other. These cover respectively aspects of living and of making a living. Thus branches of Social Geography are concerned with rural and urban settlement, and cultural attitudes that affect landscapes and ways of life, with Medical Geography, Political Geography, and Recreational Geography; while branches of Economic Geography deal with resources, agriculture, industry, transport, marketing, and with those aspects of settlement that are related to these functions.

In an integrative subject such as geography it is not surprising to find that some systematic topics overflow these neat compartments. One of these is Land Use, which is very central to geography, and is a kind of halfway house towards Regional Geography. Land, which is a resource, is affected in its use by virtually every aspect of physical and human geography, and Land Use is therefore probably the most challenging of all the systematics.

It is no longer possible for every geographer to master the whole of every systematic branch. Most universities train their students in the basic foundations of all, but thereafter allow them to specialize in one or more divisions. Because these divisions have close affinities with other sciences they play an important role as 'bridge' zones, which is of the utmost value in an age when the disruptive compartmentalization of too much specialization has led to an awareness of the need to create interdisciplinary cross-links. Geography is fortunate in possessing an extensive web of ready made cross-links not only existing within itself but also opening up into other subjects, and this means that depth of understanding in the systematics is immeasurably enhanced by a full knowledge of geography as a whole, as well as by a good grounding in the cognate subject.

Although the concept of our 'earth ball' is now increasingly familiar, the details and interrelationships of its surface are so complex that once again knowledge must be subdivided. Thus Regional as well as Systematic Geography must have branches. We cannot comprehend and understand all the features of the whole of the earth in any one instant of time, but we can describe and understand parts or sections in varying degrees of intensity. Uniformity simplifies description and comprehension: the regional geographer therefore seeks to demarcate areas wherein some similarity exists, thus distinguishing such areas from neighbouring regions. Within these regions he then studies and interprets those phenomena that give the region its distinctive character. The basis of any demarcation may be a single factor or a combination of many factors from the systematic branches. Herbertson's Natural Regions, Fleure's Human Regions, Unstead's Cultural Regions, Physiographic Regions, Functional Regions, Political Regions, City Regions, the French 'pays', Land Systems and Land Use Territories, are all attempts at a 'description explicative des paysages'. In all this, Systematic Studies are to Regional Studies as the warp is to the woof in the composite weave which constitutes geography. In this pattern of interlocking facts we begin to see the emergence of a model to crystallize our concept of geography, and we also begin to appreciate its *raison d'être* and its unique character.

This is of course the holistic approach. In the words of Professor Pye: 'Geography maintains the need for looking at things as a whole, and it is in his Regional method that the geographer makes his unique contribution to learning.' The attempt to give a total explanation of the complex relationships in any one physical and human environment is quite unique to geography, and its practice calls for a breadth and depth of knowledge rarely found in the academic world.

The scale factor is clearly important in this consideration. A small area or region will be easier to study, analyse, comprehend and describe than a large area which will inevitably involve a greater degree of physical and human diversity. As the area increases we pass rapidly from the holistic approach possible in the study of the 'pays' to the more generalized treatment inevitable in systematic geography. At the one extreme regional geography seeks to give a coherent picture of a specific place: at the other extreme systematic geography tries to give for some particular aspect a coherent picture of the world. Systematic geography therefore tends to produce general concepts, whilst regional geography reminds us continuously that nature has ordained that every place on the earth's surface shall be unique.

Our discussion so far has only taken into account the assemblage of environmental data as it exists at the present moment of time. Any attempt to provide an explanation of the present will soon reveal the need for a knowledge of the past. If we think of the warp and woof of systematic and regional geography constituting the geographic present in a two-dimensional plane, then preceding periods would be represented by other planes. Our model thus becomes three-dimensional and the depth represents time —that is, History. All of the systematic geographies have an historical or evolutionary basis and the same is true of regional geography. 'No landscape is static and even as we study what it has become it is in process of becoming something different.'

As long ago as 1756 the philosopher Kant classified empirical knowledge according to space and time:

> Description according to time is History, that according to space is Geography .... The former is a report of phenomena which follow one another and has reference to time. The latter is a report of phenomena beside each other in space. History is a narrative, Geography a description. Geography and History fill up the entire circumference of our perceptions....

We now see that geography and history can be even more intimately related than Kant envisaged, because the present geography of the world is only a momentary cross-section sliced through the infinitely greater column of the existence of world geography through time. Historical geography, which is far from being the whole of history, can be thought of as the evolutionary basis of regional geography comparable to the evolutionary bases of the systematics, already mentioned. The re-creation of past geographies at every instant of time would be an impossible task, but we can detect major periods when litttle change took place and describe the essential features that characterized them. Professor H. C. Darby's *Historical Geography of England before 1800* is an excellent example of this technique.

Any one geographical fact will fit into our three-dimensional model somewhere, since it must exist in space at some instant of time. If we could stand back and comprehend all the facts for all space and all time, then a complete understanding would be realized. This, of course, is another impossible task for the human mind. The technique, however, is applicable to restricted areas of space through selected periods of time and experienced geographers often build up an impressive volume of completeness and interconnectedness of understanding.

By now it is hoped that the nature and content of Geography together with its main aims and objectives will be a little clearer. But how are these objectives to be attained? Human intelligence, itself gradually evolving, has expressed itself in four main ways. The first which springs to our conditioned minds, but not the first to evolve, is that of written description, analysis and communication. Written communication may be either in-coming (reading) or out-going (writing) and these two skills combined may be described as literacy. Spoken communication or articulacy may similarly be in-coming (listening) or out-going (speaking). Articulacy preceded literacy in the evolutionary sequence, and whilst of considerable importance in teaching, it is a most unreliable method of communciation through time as it is dependent on human memory. To be of permanent value data must be recorded; in this book, therefore, we concern ourselves with literacy, although the place and power of articulacy as an adjunct to literacy in communication should not be forgotten.

On the model of literacy the word 'numeracy' has been devised to express human ability to communicate in numbers and other mathematical notations. Concepts and ideas which defy written description can sometimes be expressed elegantly and with brevity by mathematical symbolization. Effective communication therefore calls for the skill of numeracy as well as of literacy, and the geographer has long been aware of this desirable attribute through his interest in astronomy, surveying and map projections, which early combined to produce mathematical geography. Today the trend towards a quantitative approach in science as a whole has introduced numerical concepts, mainly in the field of statistics, into nearly all branches of geography.

Words and numbers, however, are not in themselves sufficient. Words and numbers cannot describe the layout of a new house adequately. The proud owner or designer breaks out into another form of communication. He draws a plan. The geographer cannot communicate all the finer points of the landscape completely by words or numbers alone. He draws a map. This method of visual-spatial communication has been called 'graphicacy' by analogy with literacy and numeracy. These three are the basic skills of the complete geographer, and as such we naturally devote a whole chapter to each in the first part of this book.

The three skills are not only different but also unequal. This is a consequence of a different antiquity in the evolutionary scale. Visual-spatial intelligence, which is developed as graphicacy in geography, appeared very early in our history: by the time that the brain had evolved to human levels it had acquired a marvellous inbuilt ability to operate this skill. For example, in a momentary glance at a new scene it can analyse a multiplicity of colour patterns into identifiable objects, classify them by relative position and arrive at an aesthetic evaluation. The whole process is almost instantaneous. More recently in our evolutionary story literacy has been developed to facilitate communication, and much more recently numeracy has been added to improve precision in understanding.

This slow process of evolution gives us a rather distorted impression of the relative differences of the three kinds of skill. Spatial material seems to be elementary and almost self-educating, verbal material difficult enough to need early parental instruction, and numerical material so obstruse as to warrant specialized professional teaching. However, what is really being judged is not inherent difficulty but the extent to which the brain's respective in-built abilities must be consciously supplemented. Those who find the quantitative approach in geography difficult, or who fear the mathematics, might derive some comfort from the fact that as soon as we turn from the brain to the computer the opposite picture emerges. Mathematical material now appears the simplest because it is a human artifact, whilst spatial material appears the most complex because it deals with the multitudinous dissimilar facets of the real world.

The main point for the geographer to bear in mind is that no one skill of words, numbers or maps is the simplest or is inferior or superior to the others. They are only more suitable or less suitable for particular purposes, and each may range from the very simple to the highly complex. They are complementary and not interchangeable. Each may help the others to a full comprehension and understanding of some particular situation.

To give a simple example: it may be desired to describe the layout of a village. This is a case where even a quick sketch can communicate a better impression than the most complex verbal description or the most detailed set of numerical co-ordinates. This is the case where a map simplifies not because it is physically a simpler mode of expression, but because it is the most suitable mode for that particular piece of intelligence. Each of the modes simplifies when it is the most suitable. One is aware of the pages of words that can sometimes be very economically expressed in a single mathematical formula, and conversely of very telling and significant verbal phrases that are impossible to quantify. It is the same with maps and diagrams. For example, there is one stage in meteorological understanding where numerical data spring into pattern when mapped and another stage where multi-dimensional complexity passes beyond the scope of a map and must revert to formulae.

The complete geographer must therefore aim at becoming literate, numerate and graphicate, knowing when to use each skill singly or in combination to the best advantage. Much of the necessary background training he should have received in school, but experience has shown that most students arrive at university with very different abilities in each of these three methods of communication. As these are the basic tools that all potential geographers will need to use, each receives detailed attention in the first section of this book.

The second part of our consideration then turns naturally to the collection and organization of the subject matter.

As geography is concerned with places, areas and distributions, its source material is primarily 'in the field'. A chapter on field work therefore follows, discussing data collection in both physical and human aspects of the subject. The next step is the testing of the data collected, and this forms the substance of a chapter on hypotheses and models which precedes a consideration of the systematic and regional methods of organizing and presenting the material.

The third part of our discussion turns to more practical aspects. The application of geographical knowledge to present-day problems is considered, and we conclude with a consideration of some of the decisions facing the would-be student about to launch into geography as a career. The chapter on geography in British universities will, it is hoped, provide guidance in selecting the most appropriate university departments to aim for, whilst that on careers gives an indication of the wide range of possibilities now open to those who graduate in geography.

FURTHER READING

CHORLEY, RICHARD J. and HAGGETT, PETER (eds.), *Frontiers in Geographical Teaching* (Methuen, 1965).

DICKINSON, ROBERT E., *The Makers of Modern Geography* (Routledge & Kegan Paul, 1969).

FREEMAN, T. W., *A Hundred Years of Geography* (Methuen, 1965).

HARTSHORNE, RICHARD, *The Nature of Geography* (Association of American Geographers, 1939).

HARTSHORNE, RICHARD, *Perspective on the Nature of Geography* (Association of American Geographers, 1959).

WOOLDRIDGE, S. W. and EAST, W. GORDON, *The Spirit and Purpose of Geography* (Hutchinson, 1966).

Geography and Education (Ministry of Education Pamphlet No. 39, H.M.S.O., 1960).

2

Literacy

W. R. Mead

Since geographers became geographers—and ceased to be cosmographers in the sixteenth century—they have developed three complementary languages: signs, figures and words. Of these words are the most fundamental; they can never be dispensed with. We may be upon the threshold of an age which regards words and books as a cumbersome means of recording information since more efficient methods are increasingly available for storage of facts and figures. Yet, when they are retrieved from their data banks, facts and figures still depend largely upon the use of words for their interpretation. The geographer may be a better geographer because he is numerate and employs the methods of numeracy. It may even be conceded, as one senior geographer has put it, that as the mathematization of knowledge spreads, a corresponding devaluation in the importance of literary skills may be anticipated. Ultimately, however, the geographer will not count if he is not literate.

In the broadest sense, literacy to the geographer means the same as it means to students of other disciplines. According to the *Oxford English Dictionary* it is the 'quality of being acquainted with letters'. It simply means the ability to read and write, to understand and to employ words. The purpose of this chapter is to review briefly the nature of the words used by the geographer—the vocabulary of geography; and the prose form used by the geographer—the literature of geography. It might be called an essay in geophilology—had not the word already been conceived by Frederick Rolfe for Hadrian VII in another context in the 1890s.

The vocabulary of geography

Once there were geographers, they began to assemble their vocabulary and to develop their own terminology. As geography in its early stages was largely an appendage of related sciences, so its vocabulary was originally mainly a series of borrowings from kindred disciplines. At the outset, the vocabulary of geography developed most rapidly in France. It drew its words from mariners, from the army (e.g. *plateau* and *defile*), from the literature of travel (e.g., *monsoon*), from the vernacular for natural and human features in neighbouring lands (e.g., *sierra* from Spain, *polder* and *dike* from the Netherlands), from the classics (e.g., *archipelago, meander, isthmus, delta*) and from regional dialects. François de Dainville, in *Le Langage des Géographes* (Paris, 1964), has compiled a map to show the words borrowed by academic geographers from the vernacular of various parts of France. The contributions, which number more than a hundred, are especially numerous in the south-east—the territory of alps, domes, glaciers, avalanches and chalets.

The original vocabulary of English geography—that of Germany, too—owed much to France. From Cotgrave's *Dictionary of the French and English Tongues* (London, 1611), which derived much from J. Nicot's *Thresor de la Langue françoyse* (Paris, 1606) and was one of the earliest printed dictionaries, it is possible to glean something of the sources of geographical vocabulary. The very word geography, though born in classical times, was a French importation; so, too, were hydrography, topography and globe.

In western Europe, the descriptive vocabulary of geography evolved strongly along anthropomorphic lines until the later eighteenth century. The human metaphor adopted in those days has persisted. A river has its mouth, its arms, its bed; a seacoast has its headlands, its necks of land, its nazes; a hill, its flanks and its brow; a mountain chain, its teeth. In the titles of books, for example by George Dury (*The Face of the Earth,* 1959) and A. A. Miller (*The Skin of the Earth,* 1953), the tradition is continued. When natural phenomena began to claim attention in their own right, the inadequacies of the vocabulary became evident. 'The art of describing nature is so new that words have not been invented for it,' observed Bernadin de St Pierre in his *Voyage a l'Ile de France* (1775). The deficiency was made good by substantial borrowing from other languages. Examples are found in such words as tornado (1556), volcano (1630), fiord (1674), talus (1830), drumlin (1833, earlier according to C. Embleton and Cuchlaine King), loess (1833), tundra (1841), os (more strictly ås, 1854), fjell (1860). The dates of their introduction to the English language, taken from the *Oxford English Dictionary,* are significant in themselves. As the tide of classification moved forward, waves of new words were borne in on it. Among the most evocative was the wave that sought to classify geomorphological features or rock forms by type areas—Caledonian, Silurian, Hercynian, Jurassic, Alpine, Dolomitic.

A more evolved group of words has arisen to cover concepts (e.g., isotherm, isobar, isochrone) and processes (e.g., epeirogenesis, isostasy, peneplanation, rejuvenation, nivation, congeliturbation). Sometimes words of this character are preceded by the adjectival forms of the names of those who have identified the feature or process. Examples are found in Wallace's line, the zone defined by Wallace between those areas with flora and fauna distinctive to S. E. Asia or Australia, and the Davisian cycle of erosion, named after W. M. Davis. Such phrases are of the essence of scientific terminology. All in all, at any given time, the vocabulary of the subject will be emblematic of its development. It will express the nature of thinking in the subject and, reciprocally, it will influence thought about the subject.

As one system of thinking succeeds to another, three changes take place in the vocabulary of the subject. First, a new terminology is added. Second, established words and phrases acquire new shades of meaning. Third, the general relationships between groups of words may alter because of the changing relationships between the features and concepts to which they refer. It is fashionable to define a particular system of thinking as a paradigm. Examples of changes in the general system of thinking about geography are found in the shift from anthropomorphism towards evolutionary interpretation in physical geography, in the rise of determinism and possibilism, in the emergence of the regional approach, and in the trend towards quantification and prediction. Each of these systems has its associated terminology, with key words and phrases waxing and waning responsively. When the content of the subject is in a state of flux—and such a state is inseparable from progress—the language of the subject must reflect the instability.

The problem of vocabulary was already observed in the early stages of scientific enquiry. In 1654, Christopher Wace commented that: 'The advancement of experience does necessarily propagate new words.' The origins of new words have long been a cause of dissent. They have usually forced their way into currency regardless of authority: usually, there has been no authority to admit them formally. In this respect, the situation differs between England and a number of European countries. During the later seventeenth century, the Royal Society castigated those who introduced foreign words to describe English objects. Nathaniel Fairfax, acknowledging the need of a 'well-fraught world of words that answers works', urged that any new scientific words which were required ought to be sought in the vernacular. In this context it is entertaining to compare the range of words employed to describe soils three hundred years ago with the latest English-language statement on pedology. In Volume I of the *Philosophical Transactions,* dated 1665, the 'several kinds of soils of England' were listed in a terminology which would have pleased Fairfax as 'sandy, gravilly, stony, clayie, chalky, light-mould, heathy, marsh, boggy, fenny, or cold weeping ground'. The ultimate in contemporary classification is *Soil Classification, A Comprehensive System* (U.S. Department of Agriculture, 1960). This 'seventh approximation' towards a classification divides soils into ten orders (from entisols and vertisols to oxisols and to histosols) with manifold subdivision and a careful note on the origins of the formative elements in the names employed. Such an example is symptomatic of a language explosion—though, to put the matter in its perspective, it should be recalled that even primitive cultures can be remarkably articulate in details of vital consequence to them (cf. H. Hymes, *Language in Culture and Society,* New York, 1964). Examples are found in the great ranges of words developed in the Eskimo language for snow and snow surfaces, and in the Lapp language for reindeer description.

The explosion is no less evident in the human side of the subject. The swift rise of the social sciences and the creation of an elaborate vocabulary to meet their needs has had a mounting effect upon the language of geography. The language of the social sciences tends to be peculiarly rich in abstract words and phrases. Many of these have been adopted by human geographers. Some of the words of high fashion absorbed from the social sciences have yet to find a place in the accepted dictionaries, e.g., acculturation, ekistics, parameter, interfaces. Others may have suffered distortion of meaning, e.g., sophistication, sophisticated. Abstract nouns and phrases may sometimes jostle each other so closely that they constitute prose which is near-Johnsonian in character. The vocabulary of network analysts in the field of regional science has overtones of the great man's dictionary. '*Network*—anything reticulated or decussated, at equal distance with interstices between the intersections' —the definition is his, but the language might be theirs. To introduce large numbers of abstract words in addition to specialist terminology is to run the risk of creating a private language. Geographers, like historians, have always prided themselves on being intelligible to the layman. Can they retain this ability? It is not easy to reject jargon or vogue words. Indeed, some geographers cast them into virtual incantations which recall the strings of Latin formerly incorporated by ecclesiastics in their sermons or by quacks in the description of their medicaments. Words can hypnotize. Geographers no less than other people have to battle against what Wittgenstein called the bewitchment of language.

Yet new words and phrases are necessary and the best

authors use them to clarify and not to obscure. It is not easy for publications which are heavy with abstract terminology to retain an essential clarity of thought, as is evident from such word-spinning, thought-generating books as Richard Hartshorne, *The Nature of Geography* (1939), A. R. Pred, 'Behaviour and Location', *Lund Studies in Geography,* B.27 (1967), and William Bunge, *Theoretical Geography* (Lund, 1963).

It is not surprising that with the swift growth of a geographical vocabulary and its tardy acceptance by the established dictionaries, there should have been compiled special glossaries and dictionaries of geography. In 1951, a committee had already come into being with a view to standardizing British geographical terminology. Some of its random conclusions were printed in the *Geographical Journal,* and L. D. Stamp's *A Glossary of Geographical Terms* (1961) was a direct outcome of its work. F. J. Monkhouse, in *A Dictionary of Geography* (1965), has continued assembling the vocabulary of geographers.

Nor is it possible to remain insular when dealing with geography. The subject advances in a variety of languages, and translation of texts from other tongues is exceptional rather than common. Examples are found in A. Demangeon, *The British Isles* (1939), J. H. von Thünen, *The Isolated State* (1966) and T. Hägerstrand, *The Propagation of Innovation Waves* (1967). The significance of a country's geographical output rises and falls for many reasons—not least the strategic. Accordingly, the volume of Russian geographical literature translated into West European languages is considerable. The changing reputation of a country's geographical publications is less readily explained. The giants of the German school—Emanuel Kant, Alexander von Humboldt, Karl Ritter, Friedrich Ratzel—yielded pride of place to the regionalists of France—the school of Paul Vidal de la Blache. The regional concept in turn has yielded to regional science, systems analysis and quantification, with their attendant models. The United States, Sweden and Great Britain have been the hearths of these approaches. As a result of changes in leadership, one language may succeed another in being the most important for the subject.

Today, it may be fairly claimed that it is English (or American English). Yet there is still much to be derived by all English-speaking geographers from the French language. If, as with the language of diplomacy, that of geographers is no longer French, it is still possible to assert that by reading French literature, not least the literature of geography, a great deal can be learned about writing English. The content of contemporary French geography may not always belong to the new paradigm, but the way in which it is expressed still has much to commend it.

It goes without saying that the geographer who is conscious of the importance of literacy will be prepared to have a working knowledge of more than one foreign language. As with much scientific terminology, the vocabulary of geography tends to be increasingly common to most languages but there will always be features and phenomena, especially in the field of human geography, that defy the compilers of ordinary dictionaries. Among international attempts to provide glossaries is the *Vocabularium Geographicum* (G. Quenez, Strasbourg, Council of Europe, 1967). A committee, under the chairmanship of Harald Uhlig of the University of Giessen, is seeking to clarify the European terminology of the agrarian landscape.

These, then, are some of the ways in which words used by geographers have come into being and some of the problems arising from words that perplex them. The academic geographer is faced with the dilemma of restraining the development of a private language at a time when there is a world-wide multiplication of specialist vocabulary. It does not help him that this is a problem common to most social and physical sciences. He is immediately conscious of it when he comes to work in his discipline—to contribute to or to read contributions in the literature of geography.

Style and geography

A significant literature of geography presumes a body of authors who have something to say, who have the urge to communicate it and who have the ability to do so. A flexible vocabulary which meets the needs of the subject aids communication, though there was a literature of geography before there was a distinct vocabulary of geography. But a geographical vocabulary, no matter how rich, is only one means towards communication. No less important is the way that words are used—the style that is employed. Some would argue that, for their effective presentation, different kinds of geography require different styles—that the geographer as natural scientist and the geographer as social scientist might be expected to have different ways of writing. Others would argue that style should reflect the needs of the age—that, for example, an age of functional values requires—and produces—a functional style. Certainly, there are those who have encouraged the geographer to develop a 'scientific' prose style—whatever that may be. On balance, it might be agreed that there should not be anything unique about the way that the geographer writes; though, as has been illustrated by H. C. Darby ('The Problems of Geographical Description', *Transactions I.B.G.*, 30, 1962, 1–14), the subject matter of geography can exaggerate certain aspects of the common problem of communication. Naturally, the geographer will have his particular point of view; but why—any more than the historian or the philosopher—should he wish to cultivate a deliberately different style in which to present it?

Whatever may be his feelings about these matters, the geographer should be aware that there are certain qualities of style which make for ready and appealing communication. These have been conveniently rehearsed by F. L. Lucas in his Cambridge lectures on *Style* (London, 1955) as clarity, brevity, urbanity, simplicity, sincerity, vitality—even gaiety. To the extent that geographical writing lacks any of these qualities it cannot hope to acquire the same appeal as literature in other disciplines that display them. Style is more made than born, and in the process other facts are important. First, it is usually inseparable from the regular use of the pen. Second, if he is to achieve permanent effect, the geographer will need to rewrite as well as to write. Only in so doing will he discover, like Anatole France, that a phrase long-caressed can smile in the end—and may even smile in the eyes of others. Third, it is necessary to delete with the discipline of a Robert Louis Stevenson, Rudyard Kipling or George Orwell. Fourth, it is necessary to find enjoyment in the task. Not surprisingly, a collection of essays of one of geography's most distinguished stylists—O. H. K. Spate—bears the title *Let me enjoy* (London, 1966).

More than the students of many disciplines, the geographer has the kind of training that encourages a ready pen. It is a prerequisite of his trade that he should have a facility for making notes in the field. Given this training, it should be possible for at least some geographical notebooks to approach (if not to improve upon) those of a Gilbert White or a Richard Jeffries. For most geographers, the library and the laboratory claim increasingly more time, but they do not prevent the cultivation of a lively pen. Note-making, not to be confused with note-taking, is the logical extension of work in them. Jeremy Bentham, one of the founding fathers of social science, was no model for a stylist—but he was an exemplary note-maker. It is said that he wrote with a green curtain beside him and pinned his ideas on scraps of paper to it as if they had been butterflies! Note-making helps to clarify thought: certainly it aids the organization of ideas.

While it is essential for the student to work—and play—regularly with words, he should also be aware of the way that he reads as well as of the way that he writes. All students should attempt to find out just how the eye travels over the page and under what circumstances it picks up most easily that which an author is endeavouring to communicate. As observers in the natural and human sciences and as compilers of maps and diagrams, geographers are increasingly aware of the importance of understanding the processes of perception. Perception of the printed work is a part of the mechanics of understanding. It is difficult to separate it from the content that is printed. Sometimes the pattern of black and white on the printed page and the message it is intended to convey become incandescent; sometimes they refuse to communicate. It is a useful experiment for a student to note when a page lights up—and to explain why it lights up, if he can proceed that far. Style is a critical factor in the explanation of this as well as in the propogation of ideas. It is not enough for Jean Cocteau to proclaim (*The Difficulty of Being*, London, 1966): 'I attach no importance to what people call style . . . I want to be recognized by my ideas or, better still, by the results of them.'

Literature for geographers

Ultimately, style in writing and proficiency in reading are personal matters. So, too, will be the choice of reading from the many categories of literature for geographers. The choice will blend literature from other times, other places and other subjects.

Appreciation of the subject is increased immeasurably by acquaintance with texts that have advanced it in the past as well as with those that are advancing it in the present. Examples of books which have propelled the subject forward and which marry imagination with intellect are Captain James Cook's *Journals* (especially as presented in the Hakluyt Society edition by J. C. Beaglehole, Oxford, 1955), Alexander von Humboldt, *Cosmos* (1844, 1849), Freidrich Ratzel, *Anthropogeography* (1909–12) and H. J. Mackinder, *Democratic Ideals and Reality* (London, 1919). Other times are also the concern of books which trace the history of the subject. The changing approach to the presentation of geography of other times is seen by comparing E. G. R. Taylor, *Tudor Geography 1486–1583* (1930) and Clarence Glacken, *Traces on the Rhodian Shore* (Berkeley, Calif., 1967); while evolving ideas on the physical side of the subject are included in R. J. Chorley, A. J. Dunn etc., *The History of the Study of Land Forms* (1964). In the same way as historians have their sample documents, so the contemporary geographer has a growing number of anthologies which help him to understand other lands as they were seen by early explorers and to appreciate the language that they had at their command for description and explanation. North American examples are found in John Bakeless, *The Eyes of Discovery* (New York, 1950) and

John Warkentin, *The Western Interior of Canada* (Toronto, 1964).

The appreciation of the subject is greatly increased by acquaintance with the literature of cognate disciplines. This is especially true for those who work on the frontiers of the subject and for those who are concerned with the interchange of methodologies between subjects. The choice is infinite. Take three examples—two for content, one for style. First, there has always been a close relationship between geography and history, the field of study and methodology of which is identified in E. H. Carr, *What is History?* (London, 1961). Geographers can learn much about the broader nature of their subject by substituting the word 'geography' for the word 'history' at discreet intervals throughout Carr's stimulating text. At the end of the exercise, it is legitimate to ask why there is no complementary paperback volume on *What is Geography?* Second, Lionel Robbins' succinct *Essay on the Nature and Significance of Economic Science* (London, 1935) remains a wonderfully lucid statement, the text of which can contribute to clarity of geographical thinking. Third, and in the natural sciences, J. B. S. Haldane, *Possible Worlds and other Essays* (London, 1927) is a stylistic model. It proves that in a scientific presentation, syntax need not rule out colour, piquancy—even whimsy.

It is assumed that students still find time for recreational reading. Leisure reading should make a direct as well as an indirect contribution. Those who write in the field of geography can profit from the influence of men of letters. Exponents of different branches of the subject will naturally receive different impulses from the books that they read. But common to all of them—and to put the matter at its lowest level—is the process of osmosis. To absorb subconsciously the influence of a gifted author is a positive experience in its own right. Naturally, there are certain authors who can simultaneously provide information and pleasure. The literature of travel, past as well as present, provides examples. Alexander Kinglake's *Eothen* (1864) still has passages that stand up to the masterly Near Eastern impressions of Freya Stark. Rudyard Kipling's *Letters of Travel* (1920), especially those from Canada in 1892, are keenly appreciative of the physical environment. For the geographer who would like a concentration of selected literary experiences, Margaret S. Anderson has produced an anthology of travellers' observations in *The Splendour of Earth* (1954). No home thoughts from abroad have captured the essence of any area more vividly and completely than V. Sackville-West's narrative poem *The Land* (1926). It is of the Weald contemplated from that city of poetasters, Ispahan.

The regional novel has attracted considerable geographical attention. For Great Britain, E. W. Gilbert has summarized the situation in a few succinct pages of 'The Idea of the Region' (*Geography,* XLV, 1960). His essay includes a map derived from Lucien Leclair (*Le Roman regionalist dans les Iles Britanniques,* Clermont Ferrand, 1954) which locates the territories of Britain described by 150 regional novelists. An appreciation of an individual author from a particular point of view is found in H. C. Darby, 'The Regional Geography of Thomas Hardy's Wessex' (*Geographical Review,* XXXVIII, 1948). (Hardy's Wessex is mappable and, as drawn by Emery Wallace, it is superior to any conceived by Stevenson's cartographer for *Treasure Island* or by William Faulkner for Yoknapatawpha County.) It is an entertaining experience to journey through Dorset with the Wessex novels in hand and to read their topographical passages on the spot. J. H. Paterson has conducted a parallel exercise in his paper 'Scotland through the Eyes of Sir Walter Scott' (*Scottish Geographical Magazine,* 1965, 81, 3). Scott lacked the grand panoramic sweep of Hardy, but his effect upon human appreciation of the Scottish landscape is of no less interest than that of Hardy on Wessex. John Buchan's biographical study of Scott identified the 'thunderous, cumulative topography' of the author and of its impact upon the reader, while Washington Irving, disappointed by the scenic qualities of the Border country, confessed that he saw it only through the 'magic web' thrown over it by Scott.

There is much to be gained by combining the study of other countries or areas with an intensive programme of general reading about them. This exercise has not been very widely extended. Sometimes the literature of the country is not appropriate to this form of appreciation. J. W. Watson's essay on 'Canadian Regionalism in Life and Letters' (*Geographical Journal,* 1965, 131) illustrates the surprisingly limited bibliography available in Canada. Sometimes, as with Scandinavia, it has been disregarded. Within the compass of translated literature, the range of Scandinavian literature is measured by the acutely sensitive impressionism of Johannes V. Jensen's Himmerland tales from north Jutland, by F. E. Sillanpää's powerful naturalistic novels from south-west Finland, by the hard skerry vignettes of Norway's Tarje Vesaas, by the glowing Dalarna romances of Selma Lagerlöf, and by the uncompromising realism of Iceland's Halldar Laxness. In the work of each of these authors, as in that of a large number of their compatriots, there is a keen appreciation of the natural and human scene.

The United States raises no language barrier and for the geographer it has a supremely exciting literature. It is a literature which underlines a fundamental paradox. While every endeavour has been made to create a human unity out of the physical diversity of the U.S.A., the greater part of the literature emphasizes the regional and the local. The fact was already observed more than a century ago. W. D. Howells concluded that 'our very vastness forces us into provincialism', while Merrill Jenson (*Regionalism in America,* Madison, 1951) quoted from a local newspaper of 1839: 'When we do create an original literature, it will not be general but sectional in nature ... it will spring up in nooks and corners, deriving its power and worth from . . . its circumscribed home.' In local colour, the essayist Hamlin Garland saw something even more fundamental than grass roots—'blood knowledge' of the land. Such an indigenous quality was to provide escape for American authors from being pale imitations of the European. R. L. Ramsay (*Short Stories of America,* Cambridge, 1921) extended the idea by outlining 25 'local colour' regions in American literature. In the 1930s, Howard Odum, attempting to estimate the regional distribution of fiction that might appeal to the geographer, assembled over six hundred titles. It is a reflection on the distribution of literary energy that nearly half dealt with the north-eastern states.

The 'geographical' approach of American authors varies widely. Some have deliberately set out to capture the character of local environments. H. D. Thoreau sought to extract the quintessence of the Concord area in *Walden*; Bayard Taylor, in his *Story of Kennett* (1866) endeavoured to 'copy the pastoral landscape field by field and tree by tree'. Long before Sinclair Lewis was epitomizing the small town of the Middle West in *Main Street* and *Babbitt*, J. B. Cobb found 'the miniature of the world' in Columbus, Ohio. Some authors such as Fennimore Cooper in his 'Leatherstocking' novels of the Old West, have employed extended landscape descriptions as a kind of canvas backdrop to their plots. Some have simply exuded their environment. In *Tom Sawyer* and *Huckleberry Finn*, Hannibal, Missouri, is perpetually recreated; while Mark Twain's autobiographical recollections of his steamboat days, *Life on the Mississippi*, are rich in penetrating observations on the behaviour of the Father of Rivers. The South and William Faulkner are synonymous, with a powerful environmentalist influence which is well illustrated in his description of Mississippi State (*Essays, Speeches and Public Letters*, 1967).

Much American literature, in contrast to that of Europe, has been dominated by the frontier as a fact or as a concept; though the literature of the West has been largely ephemeral. The appreciation of the natural landscape is a continuing feature in American literature—from the eighteenth-century travels of William Bartram, through the poetic reactions of Walt Whitman (for whom camp life was summed up in 'bravuras of birds, gossip of flame, clack of stick cooking my meal') to the Texan experiences of John Steinbeck in *Travels with Charlie*. It is very much the antithesis of the urban background to most American lives of 'The vascular and alive . . . language of the street' that Emerson anticipated would become the mode of literary expression.

Beside having a subconscious influence, leisure reading of fiction—and poetry—can offer an intellectual exercise. Sometimes the geographer feels overwhelmed by the number and diverse character of the facts from which he must extract significant order. It is comforting to reflect that the problem is even more complex for the man of letters. In a paper on 'The Literary Imagination and the Study of Society', delivered to Section N (Sociology) of the British Association in 1967, Richard Hoggett intimated that there was little difference between the literary imagination as it works on society and on its setting, and the mind of the social scientist endeavouring to make sense of his material. While the scientist orders his facts to make 'significant explanations . . . by conscious, controlled aggregation', the novelist orders them through imaginative power. In the final instance, it is the common spark of imaginative power that kindles science and art alike.

When the map is in tune

The language of geography is little more than several centuries old. Its vocabulary is currently experiencing an unprecedented expansion. This fact is a result of the discovery of new knowledge and of the active reappraisal of old. The vitality of the geographer's vocabulary indicates in itself that other forms of expression are only a partial substitute for words. It is paradoxical that, as substitutes for verbal expression increase, words themselves multiply faster than ever. Dictionary-makers can no longer keep pace with their multiplication and, consequently, can no longer scrutinize them and challenge their acceptability. But, if restraints on the development of a geographical vocabulary are relaxed, discipline in written style must correspondingly tighten. Beyond this, on the degree of imagination possessed by geographers and the character of the language in which it is expressed depend the life and thought of the subject.

Geography has always admitted to its ranks a broad range of practitioners. If its language and literature assume an exclusively professional character it will lose as well as gain. In so far as there is a place for those who practise on the edge of mathematics and at the margins of the abstract, there must continue to be room for those whose ideas are still best expressed in straightforward words. Nor must there be excluded from the circle a few honorary geographers who, by their word-spinning alone, have added to the literature of the subject. Geography will continue to benefit from the imaginative artist as well as from the analytical scientist. Aldous Huxley, already mentioned, must be included among the honorary company, not least for his *Literature and Science* (1963) from which the geographer may learn much about the nature of his subject. So must the essayist C. E. Montague. In the days of Copernicus and Kepler, at the dawn of modern science, there was still a music of the spheres. In *The Right Place* (1926), Montague breathed new life into the celestial analogy, communicating his own incandescent appreciation as the contours of the map created for him new harmonies that 'sing together like the Biblical stars'.

FURTHER READING

ANDERSON, MARGARET S., *Splendour of Earth* (George Philip, 1954).
GOWERS, E., *The Complete Plain Words* (H.M.S.O., 1954).
GRAVES, R. and HODGE, A., *The Reader over your Shoulder* (The Alden Press, 1947).
LUCE, A. A., *Teach Yourself Logic* (English Universities Press, 1958).
MEDAWAR, P. B., *The Art of the Soluble* (Methuen, 1967).
MONKHOUSE, F. J., *A Dictionary of Geography* (Edward Arnold, 1965).
PLUMB, J. H., *Crisis in the Humanities* (Pelican, 1964).
POTTER, SIMEON, *Our Language* (Pelican, 1950).
STAMP, L. DUDLEY, *A Glossary of Geographical Terms* (Longmans Green, 1961).
WHEWELL, W., *The Philosophy of the Inductive Sciences, Works VI* (Frank Cass, 1967).

3

Graphicacy

W. G. V. Balchin

Although literacy is the most fundamental method of intellectual communication, graphicacy is the most distinctively geographical form. Without spatial records such as maps, photographs and diagrams, geography would not be geography.

Graphicacy is the educated skill that is developed from the visual-spatial ability of intelligence, as distinct from the verbal or numerical abilities. Its development began very early in the human story. We can see this in operation today amongst the few remaining primitive peoples who have not yet reached the stage of writing; all are conscious of their environment and can 'draw' maps with the aid of twigs and pebbles. Much has been written about the map-making ability of the early eskimos, and their products compare very favourably with our best modern hydrographic charts even when covering areas of several thousand square miles. The nineteenth-century charts of the Marshall Islanders consisting of shells attached to a framework made of the mid-ribs of palm leaves are also well known and puzzled anthropologists before their use in navigation was realized. The oldest-known map is a Babylonian clay tablet from Ga Sur; this is about 4,500 years old and is now in Harvard University: many similar tablets can be seen in the British Museum. All the ancient civilizations—Egyptian, Babylonian, Greek and Chinese—were map makers and the art also developed independently in the New World among the Aztecs and Incas. There is ample information to support Professor Arthur Robinson's statement that: 'Designed graphic expression is a basic form of communication among humans. All evidence points to its existence very early in the development of man and it is known amongst most primitive people.'

Modern graphicacy, with its long history of development, has passed far beyond the simple sketch in the sand to a degree of sophistication undreamt of by early map-makers. It is no longer looked upon as an independent art—almost a curiosity—but is recognized as a fundamental support for the whole of geography, distinctive in kind but analogous in function to the fundamental supports of other subjects. Almost every subject has its own special methods of making visible what is really invisible. Thus meteorology depends on instrumentation to illuminate the invisible atmosphere, history upon documents to disinter the obscure past, and economics upon statistics to isolate data that are diffusely concealed among other aspects of daily life. Some subjects use microscopes to enlarge what is too small for the naked eye to see, but geography needs just the opposite—some form of 'macroscope' with a reducing power to scale down extensive regions to a convenient size for visual examination. Such 'macroscopes' fall into three main classes, typified respectively by the map, the photograph and the diagram.

Geographers are trained to understand the scope and potentialities of each of these three classes as well as their weaknesses and limitations. They learn how to use them correctly without misconceptions or distortions and also how to avoid being misled by the accidental or deliberate misrepresentations of others.

Maps

The first technique in this class of graphic aids is the field sketch or landscape drawing which is completely bound by the horizon and the laws of perspective. The second technique breaks the horizon bondage by adopting a higher viewpoint, thus condensing more extensive areas into panoramas or block diagrams. Third, the perspective bond is further broken by the use of still more distant and imaginary viewpoints arranged in a systematic lattice to give vertical views of all points simultaneously. This produces the map, geography's traditional tool, in which the extent of the area portrayed is limited only by the size of the paper and the scale of the microscopic reduction.

Field sketching regrettably no longer occupies as much time in geographical education as it once did, but none can deny the argument that there is no better way of becoming fully aware of the characteristic features of a piece of landscape than by sitting down and making a careful drawing of it. Learning to draw is eventually a matter of learning how to look at things, and this is precisely one of the geographer's objectives when studying a landscape. The act of drawing also facilitates the detailed analysis and study. Once again information emerges during the process of drawing or sketching which is not apparent at a first glance.

In recording and conveying information the field sketcher can also bring out particular factors by slight emphasis, he can elucidate the obscure parts of a landscape and can omit the irrelevant parts. Unlike the photographer he can be selective and can also extend his field of view indefinitely. The result is known as the drawing progresses and there is no waiting for development. Drawings can be made of landscapes which cannot be photographed because of poor light.

On the other hand a drawing is the expression of an individual draughtsman and is therefore subjective. No two drawings of a piece of landscape would be identical, and distortion could occur. Drawing takes time and time is often precious. Climatic conditions (wind, temperature, rain) are often adverse. It is not difficult therefore to see why photography has largely replaced sketching in the geographer's field armoury. Furthermore the field sketch and the block diagram are only suitable for small areas and relatively large scales. The map must be brought into use for all other cases.

The field sketch, block diagram and map all exhibit the basic characteristic of their class—they are analytical. That is to say, a high degree of cognitive analysis must go into the stage of data collection. This is the distinctive contrast between this class and the next, where data collection is carried out automatically by the camera and all analysis is reserved for the interpretation stage. The analytical nature of map data collection carries with it certain advantages and disadvantages. It is more laborious than photography but also more explicit, more selective but also more flexible. The map cannot show everything that a photograph can; it is less comprehensive. But it can also show features that the photograph cannot, such as invisible administrative boundaries or very small objects that can be represented by exaggerated symbols.

To use maps effectively, however, the geographer must be trained in cartography, and the question naturally arises as to whether the making of maps falls within his province. The answer to this question is both yes and no. In the production of basic topographic maps the geographer does not make the maps, this task being the concern of the professional surveyor: but the geographer must know how maps are made if he is to use them without error and to the best advantage. This means that the true geographer must have a background knowledge of the principles of base measurement, primary, secondary and tertiary triangulation, the measurement of height and distance, levelling, plane table mapping and map production, before he embarks on the detailed use of the maps.

The geographer may, however, need to be a map maker in the absence of suitable base materials, and this is where a background knowledge of surveying will stand him in good stead. Expeditionary mapping falls into this category. The results will doubtless be superseded in time as professional surveys become available, but until this point is reached the sketch map is better than no map. This is only like the mechanically-minded car driver carrying out his own emergency repairs. The geographer is also a map maker in that he may add, by means of field work, additional information (e.g., geomorphic, biological, land-use data, etc.) to an existing map, or he may by various comparative processes create a new map by selecting data from existing maps.

A controversial but basic element of geographical knowledge is that concerned with map projections—the problem of portraying the curved surface of the earth upon a flat sheet of paper. With large-scale maps and plans this problem can often be ignored by the geographer, but as the scale decreases so the distortion becomes more acute and it is necessary to know what inaccuracies in shape, area and distance have been introduced by the map projection employed. Position on the earth is fixed by reference to the geographical co-ordinates of latitude and longitude. The map maker's problem is the representation of the parallels and meridians of the earth's sphere on a flat piece of paper. There is almost an infinite number of mathematical solutions to this problem, but in actual practice less than twenty map projections are in common use. For the geographer the situation has been adequately summed up by A. R. Hinks:

> It is important to have a clear idea of the merits and defects of each projection, to be able to decide at once on its suitability for a given map, or when one finds it actually employed on a map to recognize what a map so constructed will do and what it will not do.

Understanding how a map is made is essential to understanding how to use it, but university departments of geography vary in the exact level of map-making expertise that they expect of the student. Some instruct all students in the fundamental aspects of map projections, field surveying and photogrammetry; others restrict one or more of these topics to special options for those with a vocational interest. There is far more unanimity, however, on the importance of cartography, the actual compilation and drawing processes that follow on from the foundational aspects.

Maps can be divided into two main types, primary and secondary. Primary maps are prepared mainly by survey methods in the field. They attempt to be a faithful representation of the real world with all its multitudinous array of shapes and forms. They do not necessarily begin with a blank page but may involve additions to existing maps, for example land-use mapping on an Ordnance Survey base.

Secondary maps are rarely prepared by data collection in the field. They are mainly derived from verbal lists, such as the distribution of car factories throughout the country, and from numerical tables such as the output of each factory. These maps are more often abstractions at least once removed from the real world.

Many maps are a compound of both primary and secondary characteristics. For example, a dot-distribution map of cattle in Britain would be chiefly secondary, but its framing coastline would be a primary feature. For every map that finally presents geographical information there may have been many more used for collecting, analysing and experimenting with the raw data. These exploratory maps are learning tools rather than demonstrating tools, and the geographer has to discard them as ruthlessly as a film editor cuts substandard shots. In map making, as in film making, there may be a high apparent wastage rate, but this is necessary to ensure that only the best goes into the final product.

While geographers accept the need to discard many of their data-processing maps they do not accept the inevitability of investing so much labour in them, and one of the most fascinating developments of contemporary geography is the investigation of ways to offload some of this work on to computers. The big challenge in this research is the fact that computers are numerate, not graphicate, and they cannot handle graphic data unless it is translated into numerical form. This means that secondary maps, which are frequently based on statistics, have proved far more susceptible to computerization than primary maps which are more purely spatial.

A great virtue of the computer is that it forces scientists to think more analytically, and therefore the development of computer graphics has been accompanied by new ways of looking at maps in order to understand their basic nature. Secondary maps, for example, are largely based on point, line or area data, and this gives rise to three subtypes: cartogram, isogram and chorogram maps, all requiring different computer-graphic techniques.

Source statistics are usually in point-data form, indicating item and quantity for each specific location. Cartogram maps quite simply and directly reproduce the numerical point-data as graphic point-data, such as dots or stylized symbols distributed at discrete locations. The computer can be programmed to identify the correct locational co-ordinates on the map and to vary the size and colour of the symbols to be plotted there. These symbols occupy only a small proportion of the map, most of which is blank space, and this means that the load of computer input-processing does not become burdensome.

Even better-suited to the computer is the isogram map which converts the numerical point-data into graphic line-data. The input work load remains the same as for cartogram maps, but the computer itself is harnessed to achieve more by calculating a close network of intermediate values in all the intervening blank spaces. Obviously this process must be restricted to types of data that lend themselves to interpretation; it has been used with great success in the field of weather forecasting. The interpolation consists of innumerable simple arithmetic calculations, a task which would be tedious and time-consuming for a human being but which is ideally suited to the high-speed computer. It prints out value symbols all over the map, and the lines of discontinuity that emerge where symbols of one value give place to those of another are the isograms, in this case isobars, that reveal the familiar patterns of weather systems. In practice these patterns are made even clearer by not printing certain values and leaving their location to stand out as conspicuous blank bands.

Thus, with isogram maps the computer is able to manufacture far more data than are orignially fed in. This creates a very favourable balance between time-consuming input processes and time-saving calculations and output processes. Isogram maps form the peak of computer-graphic achievement at the present time and lend themselves to even more sophisticated future developments such as trend surface analysis. Trend surfaces are average rather than actual distributions. Thus in a relief problem the computer is programmed to calculate the angle and direction of tilt of a plane or curved surface that could be passed through an area leaving hill summits the minimum distance above it and valley floors the minimum distance below it. This identifies the overall spatial trend and separates it from deviations from it, providing precise mathematical information about both. It can be applied not only to actual relief but also to other kinds of data that vary over space and are capable of being expressed by isograms, for example the proportion of crops to grass, the dates of diffusion of industrial techniques, and so on.

Chorogram maps are concerned with expressing numerical point-data as graphic patch-data. This is necessary when a single figure refers to an area rather than a point location. An example would be population density by parishes. This could easily be portrayed by discrete symbols positioned centrally in each parish, or by continuous isograms with layer colouring, but for some purposes it might be desired to show it in the chorogram style, i.e., patches of shading extending right up to the boundaries of the relevant parishes. Chorogram maps are far less easy to computerize because it is no longer sufficient to input a single set of locational co-ordinates for each population value. Instead there must be a larger number of co-ordinates to define the irregular outline of each parish, recognizably if not precisely. This increases the work load of the input processing so much that chorogram maps are often avoided in computer-graphic practice. An attempt is being made to overcome this problem by substituting regular grid squares for irregular administrative areas, as square patches can be rapidly computerized by formula. In order to capitalize this modification a number of official data-collecting agencies, such as the population census, are preparing to maintain their records on a grid square basis, and so gridded chorogram maps are likely to be more in evidence in the future.

While secondary maps, as abstractions, have lent themselves to computer-graphics, primary maps, concerned with the complexities of real shapes and forms, still present major problems for the computer. Since they are primary they are themselves the starting point, and the computer is required not to make them but to read or analyse them. This is a far less mechanical process, involving recognition, analysis and judgement, which we are still a long way from knowing how to programme. We still need to develop non-computerized map-reading techniques to a higher pitch of analytical understanding before the computer barrier can be broken for primary and morphogram maps.

Map reading can elucidate why such a barrier exists. It is due to a fundamental difference between graphicate communication and literate and numerate communication. The first is concerned with pattern and the last two with sequence. Words are always read in the same fixed line sequence and numbers must use either a row sequence or a column sequence, but we should not progress very far in understanding a map or picture if we started at the top left-hand corner and worked systematically, point by point, down to the bottom right. This is what a scanner does, with the result that it is no more than a copying device. It can cause a television picture to appear, but it leaves it to the viewer to interpret.

Pattern recognition, the key to understanding spatial relationships, is appreciated in a definitively non-sequential way. The eye roves at random over the whole map seeking out some dominant pattern to use as a framework for relating lesser detail. The objective is to bring to light patterns, distributions and features 'hidden' in the complexity of the information displayed, also to seek correlation between otherwise disparate factors such as physiography, soils, vegetation, climate, settlement patterns, etc. Distribution patterns and correlations which cannot be deduced from field work and ground study alone will emerge from the bird's eye view which the map gives to the analytical geographer. In the words of the American geographer Preston James:

> To a very small creature living on the surface of a half-tone photograph the detail of the printed dots would become quite familiar ... yet the larger pattern of its dots which are combined in the general areas of light and shade to form the lineaments of a picture would not really be at all obvious.

The analysis and interpretation of the Bournemouth sheet of the Second Land Utilisation Survey may be cited as an example. This map contains two quite distinct sets of data, the Ordnance Survey base which includes contours, streams, place-names, etc., and the Land Utilisation overprinting which distinguishes a great many functional characteristics of the land. To the untrained eye this map presents itself as a mass of coloured patches, but the geographer can order them into significant groupings capable of correlation with features of the terrain. He first sees a remarkably consistent correlation between unbroken stretches of pasture and the flood plains of the three chief rivers. Then he compares meaningful mixtures of uses constituting farmscape on the low terraces and wildscape on the higher plateaux. He also notes that on the coastal section of both terraces and plateaux, farmscape and wildscape have been superseded by townscape in the coastal resorts of Bournemouth, Poole and Christchurch. He may also realize, if his cartographic insight has been exceptionally well trained, that the terrace farmscape is oriented towards supplying the towns with perishable foodstuffs.

The Bournemouth example illustrates the highly individual nature of patterns. Experience with the Bournemouth sheet may prove helpful when tackling another land-use map, or it may prove completely irrelevant. For this reason geographical training offers wide experience of a great variety of maps in order to give the student a well-stocked store of analogies to draw upon when he encounters new areas.

A modern development is the streamlining of this 'massive experience' approach by introducing more and more systematic guidelines. The reading of contours is an example with which the student is already familiar because he has been early exposed to a set of systematic exercises that guarantee his subsequent recognition of any form that contours can express. One by one other features of maps are also being isolated and systematized. The land-use student, for example, no longer looks for immediate correlation with terrain. Instead he establishes the existence of purely land-use patterns, in their own right, according to a set of operational guidelines, and only subsequently does he correlate them with other distributions, such as soil or rainfall, which have also been independently established.

In this trend towards systematization, quantification plays a large and increasing role. Quantification, as is explained in the next chapter, 'Numeracy', allows for more precise and meaningful comparisons to be made. It is only moderately enlightening to identify an area as consisting of marginal farmland with a scatter of idle land, but far more informative to say that idle land occupies 30 per cent of the area and occurs in patches at an average distance of 400 yards apart. The quantitative exploration of patterns on the map is a very active growing point in geography today and new systematic methods are developing in rich variety. Cole and King's *Quantitative Geography* can be recommended in this context as a book which ties quantification firmly to a spatial basis.

Photographs

The subgroups of photographic aids have a close correspondence to those of cartographic aids. The analogue of the field sketch is the ground photo, that of the block diagram the oblique air photo, and that of the map the vertical air photo.

All of these are primary documents produced by direct spatial methods. They are also morphogram documents that faithfully reflect the real world. They are complete and correct records of what was visible at the instant they were taken, adding nothing, subtracting nothing and changing nothing, but by the same token contributing nothing to their interpretation. The interpreter is not helped by an inbuilt analysis as he is with maps.

The surveyor was quick to realize the potential value of photography as an aid to map making. The fact that it is an instantaneous operation, capable of being carried out from a moving vehicle, vessel or aircraft has meant that the analytical side of the work could be removed from the arduous conditions of the field to the more convenient conditions of the drawing office. A Frenchman, Laussedalt, is credited with the first ground photographic survey as early as 1858, but the principles were largely worked out by British surveyors in the Himalayas and Swiss surveyors in the

Alps towards the end of the nineteenth century. Air photography also came early and actually preceded the aeroplane, as kites and balloons were used from 1856. During the war of 1871 the French used aerial photographs for intelligence purposes, and both the First and Second World Wars gave a great impetus to the application of photography to surveying. These activities have given rise to the separate science of photogrammetry. A modern development is the photo-map which retains the photograph as a base and superimposes cartographic detail such as place names and colour washes. Symbols may also be used for features too small to be seen at the scale of reproduction. In skilled hands the photo-map can sometimes be made more effective than the conventional map. The moon has lent itself extraordinarily well to this technique.

The end product of photo-interpretation is not necessarily a map. It may be a verbal report or a table of statistics. But in all cases it is a morphogram document that is being interpreted, and this means that pattern recognition skill cannot be dispensed with. Furthermore, photographic patterns may be less consistent than map patterns. Whereas on the map a lake would always be the same unequivocally-keyed blue, it might well appear as a dark patch on a morning photograph but quite bright in the afternoon, and patchily patterned by mountain shadows in the evening. The interpreter has to distinguish between the permanent and the temporary characteristics of any particular feature. The fascinating story of how air photo interpretation techniques were extended and developed during the Second World War is contained in Constance Babington Smith's *Evidence in Camera* (Penguin, 1957), an excellent background book for intending geography students.

The same book reveals that German photographic intelligence in the Second World War was vastly inferior to British because the Germans relied on single prints whereas the British used stereoscopic pairs. Single prints are perfectly adequate for ground and oblique air photos which obey the familiar laws of perspective, but verticals are difficult since we see a strange world not easy to interpret. Mountains and hills are flattened and may apparently disappear, while soil and vegetation patterns which we cannot see on the ground may be emphasized. Understanding can be accelerated by the use of stereoscopic lenses that can fuse two overlapping photographs into a single three dimensional image. Stereoscopic interpretation is now an essential ingredient in the geographer's training.

In recent years several other techniques have been developed to make air photographs more definite. Colour photography is the most obvious of these and there has also been investigation of picture production with infra-red, radar, and other wavelengths. Each of these gives a more unequivocal appearance to one or more features that are indistinct or even invisible in ordinary light, and their use in combination will ultimately help to automate photographic pattern recognition.

Mention must also be made of photography by satellite. Rapid mapping of a large area of the earth's surface on a small scale has already been demonstrated in a photomosaic map of Peru prepared from Gemini photographs by the United States Geological Survey. This map covers nearly a third of a million square miles, yet the data-collection process occupied less than three minutes. It has been suggested that space photography would also be highly appropriate for world land-use mapping, and that in cloudless conditions the data could be collected in six days from a satellite at a height of 125 miles. However, a great deal of pattern-recognition research remains to be done before the resultant photographs can be translated into more than the most generalized land-use categories. Resolution of detail is likely to be greatly impaired by the need to transmit data instead of actually retrieving film; the difference in quality can be testified by all who have compared the transmissions from the moon with pictures reproduced from film brought back to earth.

One other class of photographic aid must also be mentioned. Motion pictures can be of the greatest value in the understanding of dynamic processes that are either too fast or too slow for proper analysis. Rapid events can be projected in slow motion and stopped at critical points for more intensive study. Examples include the mechanics of a breaking wave, turbulent flow in rivers and the flow of pedestrians or traffic. At the other end of the scale extremely slow processes can be speeded up to elucidate the nature of the change. This has long been a popular technique for recording the growth of vegetation and has also yielded some surprises in other fields. During the International Geophysical Year of 1957-8, vertical photographs of the aurora borealis, taken with a time-lapse camera pointing upward, revealed a clear directional trend in the aurora's apprarently aimless movements, and also that the direction reverses itself at midnight. More recently the speeding-up of cloud pictures taken from satellites has revealed the existence of wave clouds moving radially, a phenomenon that was previously unsuspected.

Although animated cartography is also possible, it cannot have the same value as cinéphotography. Because of cartography's analytical nature, the conclusions would have to be known before the film could be prepared. Its research uses would therefore be extremely limited and its main function would be too demonstrative. For this reason the time aspect in relation to maps is more likely to be pursued with the aid of special diagrams—which are discussed in the next section—rather than with motion sequences.

Diagrams

The geographer turns from maps to diagrams when he wishes to study relationships other than the spatial distribution of phenomena over the earth's surface. If, for example, he wishes to dissect out vertical relationships, he uses sections; for time relationships he uses graphs or dispersion diagrams. Vector diagrams, such as star graphs or clock graphs, are employed when directions are to be compared, and scalar diagrams when the comparison is between non-directional quantities. These latter may take on a variety of shapes, including pyramids to show age and sex structure, columnar diagrams or bar graphs to show types of trade or other simple quantities, divided circles or pie-graphs to show quantities as a proportion of unity, and so on. Logarithmic graphs are introduced to elucidate rates of change, and flow diagrams when movement is being studied.

The upsurge of interest in inferential statistics has placed a great deal of emphasis upon histograms and scattergrams which arrange data in forms suitable for testing. Histograms arrange the statistics for single forms of frequency of occurrence along a quantitative scale, and are of value for showing how closely the distribution approaches the normal curve. Scattergrams arrange the statistics in relation to cartesian co-ordinates and are of value for showing how closely they approximate to a line average, or regression line.

Other important types of diagram are those that have predictive value, for example time series and their derivates, such as moving averages, and those related to organization and decision making, such as network analysis and linear programming techniques.

Although the three classes of graphic aids have been discussed separately, in practice they show a substantial measure of overlap. Diagrams, for example, frequently function as cartograms, showing the distribution of relationships across the face of a map. The photo-map is another case of overlap that has already been cited, and doubtless in time diagrams will also be adapted to form cartograms on a photo-base. It is certain that the overlap will continue to increase as geographers seek additional ways of combining the light shed on special relationships by diagrams with the speed and comprehensiveness of photographs and the analytical explicitness of maps.

FURTHER READING

BABINGTON SMITH, C., *Evidence in Camera* (Penguin, 1957).
COLE, J. P. and KING, C. A. M., *Quantitative Geography* (John Wiley, 1968).
DICKINSON, G. C., *Statistical Mapping and the Presentation of Statistics* (Edward Arnold, 1963).
HUTCHINGS, GEOFFREY E., *Landscape Drawing* (Methuen, 1960).
MONKHOUSE, F. J. and WILKINSON, H. R., *Maps and Diagrams* (Methuen, 1963).

4

Numeracy

S. Gregory

The need for geographers to be numerate as well as literate is no new development. Indeed, the accurate location of places has been a traditional geographical preoccupation, and the intellectual and practical skills of astronomy, surveying and map projections loomed large in all early treatises on the subject. Moreover, many of these aspects of 'mathematical geography' are still of major relevance today, especially those concerned with field surveying and its extension into the realms of photogrammetry and air photo interpretation, i.e., those fields considered in the preceding chapter, 'Graphicacy'. Other forms of simple numerical manipulation have also long been used by geographers; the basic data of climatology, economic geography and population studies, for example, are often of a numerical nature.

Over the last ten to fifteen years, however, new trends have appeared in the subject, involving the more systematic and organized application of mathematical reasoning to geographical problems. These trends consist partly of a more sophisticated handling of the traditional type of numerical data, using more complex techniques of analysis and synthesis, and partly of the formalizing of known or assumed geographical relationships into general systems that are often capable of numerical or quantitative evaluation. In contrast to those surviving elements of traditional mathematical geography which rely largely on Euclidean geometry and associated trigonometry, these newer developments rely heavily on statistical theory together with many aspects of both pure and applied mathematics.

This trend should not be a cause for alarm to the intending student, even if he considers himself essentially non-mathematical. The full rigour of mathematical technique appears only in more advanced geographical research and theoretical studies. At undergraduate level, and indeed in a large proportion of geographical work, the requisite mathematical and statistical understanding is no more than can be reasonably expected from any educated individual in the contemporary world. If this is more than the student feels he possesses he will not be alone in this situation. Departments of Geography realize that their students are likely to come to university more graphicate than numerate and are prepared to supply the necessary background. In the future, as society grows increasingly numerate, today's geography students will be glad to have had this opportunity.

Traditionally, the basic tool of the geographer has been the map, sometimes used simply to locate places described in the text, and sometimes used to show the distribution of relevant phenomena. In these ways the map has become the basic data-processing and data-presenting technique, utilized to gain an understanding of the geographical *where?*, *what?*, and *why?* Yet for none of these is the map invariably the most satisfactory means to achieve the goal of geographical understanding.

In the definition of *where?*, the may clearly has a great deal to commend it, but the limitations imposed by scale must always be borne in mind, while the representation of a three-dimensional earth surface on a two-dimensional plane of paper means that the characteristics of altitudinal location are always inadequately portrayed. There is also the perennial difficulty of transforming the curved surface of the globe on to a flat surface, a difficulty resolved by mathematical transformations of spatial co-ordinates via map projections. But despite these limitations in the accurate representation of *where?*, the map obviously does provide a valuable, legitimate and indispensible means of presenting spatial characteristics. The numerate alternative of specifying co-ordinates in two-, three, or n-dimensional space is only of full value in completely computer-oriented investigations.

As a means of presenting information to answer the geographical *what?*, the map is again of great value. It is at its most useful either for simple, observable phenomena, each item of which is large enough to be represented at the scale used, or for derived characteristics which have a virtually continuous or semi-continuous distribution. In this way, patterns of land-use, mean rainfall, vegetation or soil groups can be shown. Again, derived values such as total production values, densities or variability characteristics can be presented for pre-determined units such as administrative areas. However, for certain detailed studies such as till fabric analyses, farm studies, or many socio-geographic investigations, the mapping of the end product is far less satisfactory. This type of study requires some form of sampling as the basis for data collection, a theme that will be taken up later, while the map tends to require total or near-total enumeration of the phenomenon under study. It is true that the sample results can be used to estimate total conditions, but they nevertheless make mapping less easy and less satisfactory. In many ways, the geographer's pre-occupation with mapping his end results has tended to condition the sort of problem that he has investigated and the type of data that have been collected, whilst the focus on total enumeration to facilitate mapping has often influenced both the physical scope and the intellectual content of his enquiry.

It is when the question *why?* is to be answered, i.e., when the cause of things is being examined, that the map becomes an increasingly inadequate device. The more complex the problem happens to be, the less useful is the map as the basis for explanation. The interpretation of mapped data in a causal sense normally relies upon subjective judgement and experience, drawing on knowledge and understanding not presented on the map. This literary-cum-historical approach based on personal interpretation is one of considerable standing, and the results obtained are often perfectly valid. It is, however, very difficult if not impossible to establish facts firmly, for there are no agreed and accepted criteria which can be used. The attempt to explain one mapped distribution by the mapped distributions of some half-adozen partial causes, whether by the use of overlays or any other visual method, could—in extreme cases—lead to as many interpretations as there are interpreters, with no means of deciding which solution, if any, was the correct one. Yet the ability to test the validity of one's conclusions, and to establish which of several explanations is the most likely to be correct, is obviously not only desirable but also essential.

Science, in general, aims to establish validity through experimental investigation and confirmation, but in field sciences such as geography this is often not a feasible proposition. Informal (if informed) guesswork cannot be replaced by experimental proof, and instead it must be replaced by the testing of hypotheses in terms of spatial probability, and by the mathematical manipulation of data into conceptual frameworks that allow such testing to be carried out.

A growth in geographical numeracy is thus essential, partly to answer the question *what?* in certain cases, and partly to tackle the question *why?* without reliance upon unassisted subjective interpretation. One could well ask, however, why this need has been appreciated only now when it has obviously existed in the past. To some extent it results from the fact that only in the last decade and a half have desk calculators and electronic computers become more widely available, providing the material equipment to permit and encourage this growth in quantitative techniques. However, there are other, and perhaps more fundamental, reasons which have also contributed. What has been termed the 'data explosion' has resulted in the technical availability of so much information that it is impossible to use even a part of the relative data without one or more of the following facilities: automated handling, sifting and analysing by computer and the use of sampling procedures rather than total enumeration. These imply the need to use and understand a range of mathematical and statistical procedures as *sine qua non* for satisfactory data utilization.

Another underlying force is represented by the growing tendency for geographers to become involved in planning issues, whether as consultants, as professional planners, or as research workers whose investigations are aimed at some practical application. In all such fields, the results of an investigation are only of real value if the degree of validity of the findings can be clearly specified and if they are presented in such a way that policy decisions can be based on them. Both of these require findings to be presented in quantitative terms, so that decision-making processes can be assisted. This example is really but a specific illustration of a general case. As geographers become involved in inter-disciplinary studies, which require their ideas and results to be communicated to, and understood by, workers from other fields of study, it is essential that a common mode of approach be used to assist the exchange of ideas and the integration of results. This common mode is increasingly dependent on a mutual base of numeracy, for the statistical techniques now being widely employed in geography have been applied in most cognate fields of scientific and social scientific enquiry over a much larger period. Geography is thus peculiar not in its adoption of these techniques and concepts, but rather in the belatedness of its move in this direction. In doing so, it must perforce lose its role as one of the major repositories for refugees from mathematics!

By its moves in this direction, geography is in fact fitting in with a basic trend in contemporary scientific philosophy. Thus Philip Morrison (in *Adventures of the Mind,* 1963) states: 'The rise of statistical prediction, of probability, is perhaps the most characteristic of all the developments of twentieth-century science.' For geography, following this path eliminates the need for determinism, and allows probabilism to discard its subjective guesswork. In this way, with the statistical, probabilistic approach being adapted to the analysis of geographical problems, the geographer begins to think in terms of the 'definite maybe'! The other aspect of contemporary trends that needs stressing is expressed by Stephen Toulmin (in *The Philosophy of Science,* 1953); 'Only when a regularity has already been recognized or suspected can the planning of an experiment begin: until that time the mere multiplication of experiments is comparatively fruitless . . . and the accumulation of observations in large numbers will be as much a waste of energy in physics as in cartography.' This clearly implies that the theory or hypothesis is primary, and that experiments or investigations are carried out to test its validity. Such testing must needs be as convincing as possible: the theory or hypothesis must therefore be stated unambiguously, and in such a way that results can be objectively compared with it. This necessitates the expression of both theory and observations in numerical terms, as far as possible, and the use of accepted statistical techniques as testing methods. The relevance of such an approach to geography is discussed in a later chapter under the heading of 'Models'; not all of these involve the need for numeracy in a direct sense, but their application to the real world in any useful way necessarily involves expressing conditions in quantitative terms.

As has been indicated earlier, the use of sampling techniques as the basis of data collection is being increasingly appreciated in geography. In the contemporary world such a development should be accepted as axiomatic, for we live in a society which relies on sample results to provide much of its information. Political attitudes and opinions of millions of people are estimated from samples of but a few thousand, and the results are blazoned across the nation's newspapers. Rival television programmes are compared in terms of their viewing audience, and that which attracts only 3 to 4 million viewers whilst its competitor is seen by 10 to 12 million may soon cease to exist—yet these figures again may be estimates based on a sample of only a few thousand individuals. Advertising and marketing campaigns, national and local government policies, planning decisions and industrial production control—all these and many more depend essentially on data obtained by sampling procedures.

It is the final phrase—*sampling procedures*—which is important. Sampling does not imply simply picking a few items—*any* items—rather than taking all items; it implies definite procedures by which those items to be picked are decided. The purposes of such procedures include the attempts to ensure that the sample chosen is adequately representative of the total body of data (referred to as the data population) and that no bias, deliberate or accidental, is introduced into the selection process. Unfortunately, the term 'sample' has previously been used very loosely in much geographical writing, for many so-called 'sample studies' are in fact 'case studies'. Thus, a number of cases are taken, selected because they are known or thought to be good examples, and these are studied to illustrate certain conditions; at best, these could be called purposive samples, selected deliberately for a particular purpose. They are thus biased from the very beginning. Bias can also be introduced in many other ways. When selecting farms for study, for example, there is every temptation to restrict one's choice to those with relatively easy access, or to those where the farmer is helpful and co-operative (or where the farmer's wife can be relied upon to provide tea!). Again when selecting shoppers or commuters for direct personal enquiry in an urban geographical survey, the natural tendency may well be to choose pleasant rather than surly-looking people to question.

Unless these and other less obvious tendencies towards bias are avoided, it is not possible to use the sample data as an adequate basis from which to estimate the characteristics of the total data population. To be able to do this, sampling procedures must ensure that the selection of sample items is strictly random, i.e., that each item from the total body of items has an equal chance of being selected as part of the sample. This randomness is ensured either by the use of prepared tables containing numbers occurring at random, or by programming a computer to print out such a list of random numbers. The selection of sample items is thus determined without bias. There are, of course, innumerable modifications of various degrees of sophistication that can be employed. A systematic sample, i.e., items at some regular interval in time or space, can be used, provided that it can be shown that bias will not result, while it is usually desirable to divide the total population into a series of subgroups or strata on some logical basis, and sample each stratum separately.

Whichever detailed sampling procedure is used, however, certain advantages accrue as compared to the use of total enumeration. Given that only a certain amount of time is available for a particular investigation, whether this be on a school or university field week or on a large research project, sampling can mean *either* that a larger area or theme can be considered *or* that the same area or theme can be studied in greater detail. Conversely, sampling also implies that any particular problem can be satisfactorily considered in a shorter period of time. This ability to adjust project and time, together with the possibilities of quality control in data collection, are important practical considerations. Equally important is the fact that, provided the sampling procedure is sound, the results obtained from the sample study can be used as a valid estimate of conditions over the whole population. It is this function which underlies the whole sampling approach, and it is to effect this that sound statistical procedures must be employed both in selecting a sample and in processing the results afterwards.

This type of approach can and should be employed in all fields of geographical enquiry. In geomorphology, the contemporary shift of interest away from descriptive evolutionary accounts of areas and towards problems of process, both permits and requires sampling techniques. The character and mechanisms of slope form and movement, of beach composition and development, of river-channel shape, slope and load, all need careful and detailed field measurement which can only be effected at sample sites. Total enumeration is not possible, for the number of potential measurement sites is infinite. The same applies to studies of glacial deposits and soils, of vegetation patterns and composition, and local- or micro-climatic phenomena. Equally, studies involving people and their activities, whether these be social or economic, often require direct personal enquiry, and if the problem is on any scale other than the very smallest, then a sample of individuals becomes essential. Similarly, it is samples of shops, transport facilities and other services, or of industrial and business premises, that must often be used, rather than incorporating all such organizations within an area of study, for this both permits a large problem to be investigated and gives sufficient time to effect adequate control of data quality.

As stressed earlier, however, the purpose of such sampling is to obtain sufficient reliable data from which to estimate the characteristics of the total population. The word 'estimate' is used advisedly, for no sample result is likely to give exactly the right answer for the total population. What is possible, however, is to calculate the margin of error appropriate to the sample result. Thus the value for total conditions can be presented by means of the sample value, plus or minus a specific sampling error. In this way limits can be presented within which the true value is likely to lie, and also the percentage probability of this being correct can be assessed. It is this element of imprecision, this inability to present a clear-cut, one hundred per cent accurate, categorical statement, which offends or perturbs many geographers unaccustomed to working in probabilistic rather than deterministic terms. Yet the degrees of probability normally insisted upon are reasonably stringent. Results have at least a 95 per cent probability of being correct before they are acceptable at minimum levels, and probabilities of 99 per cent or even 99.9 per cent are often looked upon as being desirable, and perhaps essential. Such limits may well be more demanding than the degree of accuracy of observation possible when a large problem is tackled by total enumeration rather than by sampling; certainly this has been suggested for many sample censuses that have been taken.

The existence of sampling error must also be borne in mind when two or more sets of sample data are compared. This is effected by means of a wide choice of possible tests which attempt to evaluate the *statistical significance* of any apparent similarities or differences between such sets of sample values. These tests of statistical significance provide objective methods in order to achieve the basic geographical aim of evaluating the degree of relationship between apparently associated phenomena, but for true geographical as well as statistical significance it is necessary to demonstrate a logical as well as numerical link. Furthermore, objective assessment need not be restricted to such obviously quantitative data as climatic characteristics, production figures, population or locational co-ordinates. There also exist appropriate tests for assessing relationships between *ordinal* or ranked data and even between *nominal* data such as settlements of different historical origin, deposits of different lithology or people of different social or cultural affinities.

As more complex problems are investigated (and the real world with which the geographer is concerned is nothing if not complex), so techniques of greater sophistication become both necessary and available. The geographical need to assess the individual and inter-related contributions of a whole host of partial causes poses a continuing problem. Related, and perhaps even more difficult, problems are also posed by the demands of synthesis and generalization, and by the need for logical classificatory techniques of a taxonomic nature for such purposes as regionalization and conceptual model making. Such tasks may be first attempted in terms of personal subjective judgement, but subsequently they must be validated by more objective quantitative techniques. These techniques all provide invaluable tools by which to effect more satisfactory solutions to these perennial geographical problems, and the undergraduate will probably be introduced to analysis of variance, factorial design, partial and multiple correlation and regression, factor analysis and principal components analysis.

Clearly, if geographers are to get maximum returns from quantitative techniques, it is essential that the underlying concepts are fully understood. This is especially true of the basic frequency distributions upon which so much statistical theory rests. Thus, most techniques assume that the total population data fit what is termed the 'normal curve', and from this are derived many of the elements of sampling theory and sampling errors, the basic tests of statistical significance, assessments of probability, and many of the more advanced techniques mentioned above. It is therefore essential to test for normality before using such methods, and if non-normal data are found then either to use other tests and techniques or to create normality by the use of mathematical transformation such as logarithms, roots, or powers. It is also necessary to know of the numerous non-normal distributions that underlie the field of probability theory, since these form the basis of many studies of distribution patterns, so fundamental a part of geography.

Thus, from the simple samples, comparisons and models

appropriate to school geography can be evolved and incorporated an ever-expanding body of methods which allow geographical research upon increasingly complex problems. The largely non-numerate type of training that has characterized geography in the past has now been modified as a result of the need to handle quantitative techniques, whether for such statistical analyses as those mentioned above or for the formulation and utilization of mathematical models. The gradual growth of the 'new mathematics' in schools, which may involve the use of set theory from the age of six and matrix algebra from the age of twelve, will thus present new possibilities for the integration of numeracy more fully into geography.

Other possibilities are also raised by the growing availability of equipment for the handling of numerical data. Most geographers in the future will be accustomed to working with electric or electronic calculators, which remove the drudgery from computational work once practice has led to operational efficiency. At the research level, the electronic computer is equally becoming a normally accepted tool for geographical work. The ability to handle large bodies of data, and to carry out both complex calculations and repetitive analysis, whether this be complex or simple, opens vast new possibilities in terms of the scope of the geographical problems that can be investigated. Additionally, of course, there is the use of the computer for basic graph and map construction. The integration of these two functions means that from raw data to map, if map be needed, can be but one step—a detailed computer programme.

Thus, for the geographer of the future—and, indeed, of the present—numeracy is a fundamental requirement. It does not, however, *replace* the pre-existing needs for literacy and graphicacy, nor for a thorough understanding of his field of study. Rather it complements these, adding a further dimension to the geographer's training, a further tool to his collection of techniques, and further refinement and precision to his processes of rational thought.

FURTHER READING

COLE, J. P. and KING, C. A. M., *Quantitative Geography* (John Wiley, 1968).
GREGORY, S., *Statistical Methods and the Geographer* (Longmans Green, 1968).
LANGLEY, R., *Practical Statistics* (Pan Books, 1968).
MORONEY, M. J., *Facts from Figures* (Pelican, 1956).

5

Field Work

Alice Coleman

Field work is the geographer's planned opportunity to experience the raw material of his subject. His subject is 'place' and his raw material 'places'. By studying individual places at first hand in the field he becomes better equipped to understand places studied at second hand in books and maps, and hence, gradually, to construct the complex web of concepts that constitutes insight into place in general.

Because it is the means of first-hand access to raw material, field work is absolutely basic to geography. Indeed, one way of writing a history of geography would be by tracing the development of field studies. This development can be likened to a six-stranded rope threaded through the centuries, each strand thickening or thinning over time with the waxing or waning of the element it represents.

Units

The earliest emphasis was on the *exploratory* strand, laying a premium on the collection of any and every kind of place material. This stressed the cognition of objects, or 'units' in the perception sense, and was completely unselective. It was not possible to sift the wheat from the chaff at this stage; all was grist to the mill, curiosities and travellers' tales equally with what subsequently proved to be of scientific value.

This strand is characteristic of the early stages of each science and also of each individual, and it declines in importance as the ratio of the known to the unknown increases. But it can never be entirely dispensed with. Whenever the individual visits a new environment, and whenever a science pioneers a new frontier, exploration resumes its importance. This has been recently illustrated in the case of geology. Confronted with an entirely new environment, the moon, it has resorted to magpie-like collecting of any and every kind of rock.

The value of the exploratory strand is in its open-mindedness. It does not ignore facts simply because they lie outside established conceptual frameworks. It is geared to contexts of ignorance, in which undirected fact finding can still play a useful role as a preliminary scanning operation.

From the viewpoint of the student each of the six strands involves both demonstration and discovery, in which he is respectively recipient and participant. In the case of the exploratory strand the usefulness of demonstration is still accepted, especially in highly unfamiliar environments, but discovery is tending to be phased out of undergraduate field programmes in favour of the five remaining strands where thinking is more directed and student participation yields better educational value per unit of time invested. Discovery within the purely exploratory context is being left more to the original research worker.

Of itself, exploratory investigation is not an indefinitely self-sustaining activity. Ultimately it reaches a point at which the accumulation of material exceeds the capacity of the human memory and is perceived only as a confusion of multitudinous assorted facts. In order to regain mental control over them a different and more directed type of thinking must be introduced. The facts must be analysed and organized into some kind of pattern or framework which can act as a filing system for knowledge and so greatly facilitate its assimilation. All of the five remaining strands of field work are concerned with some kind of spatial pattern making.

Classes

The second strand represents the simplest kind of pattern making—*classification*. The unit facts are analysed for contrasts and correspondences, and are grouped into categories accordingly. These categories may be either systematic, if based primarily on content, or regional, if based primarily on distribution. Both are implicit in this second strand.

Geographical classification has often been stimulated in the first instance by the more obvious and extreme types of contrast, namely those which exist on a planetary or continental scale. Thus the first climatic classification was the ancient Greek division of the globe into torrid, frigid and temperate zones, early zoo-geographical work was inspired by the discovery of exotic animals in Latin America, and soil classification began in Russia, a country large enough to contain marked contrasts within itself. Gradually, however, the development of more precise mapping and measurement has paved the way to the recognition of progressively more subtle and detailed systematic and regional differentiation.

The decisive step in this direction was taken by Alexander von Humboldt (1769-1859), whose field work first combined measurement and mapping as a basis for classification. His isotherm and other maps of South America opened geographers' eyes to a vast range of possibilities and gave the subject its modern scientific status.

In the exploratory strand the development of the individual recapitulates that of the science, but in the classificatory strand the reverse is true. The science has evolved from the exotic to the familiar; the student studies the familiar first and works from the known to the unknown. Therefore most university geography departments try to include at least one foreign visit in their undergraduate field programmes.

Progress from unit to class in field work is progress from percept to concept. This has both advantages and disadvantages. The advantage of a classification is its structure which provides context for its component concepts. Once mastered, this context becomes implicit. It does not have to be rehearsed explicitly on every occasion that the individual concept is invoked, and consequently the stream of thought is much less cluttered. Consider how thinking would be impeded if a concept such as 'shopping centre' had always to be considered in explicit context. The student would first have to take cognizance of a bewildering array of unit objects, then classify them into bricks, glass, foodstuffs, etc., then classify these into more complex classes such as walls, windows or stock, and so on, through several more levels of abstraction. Fortunately this is not necessary. A well-established concept can often be observed as simply, directly and rapidly as a percept. This is a testimony to the tremendous value of words in crystallizing concepts and streamlining thought.

However, this advantage can become a disadvantage if the classification is faulty, because concepts may be observed without seeing that the underlying percepts are at variance with them. Observation is no longer open-minded but channelled, and sometimes blinkered, by a man-made conceptual framework. The pattern may become so dominant that the facts become recessive. This is a phenomenon that has long been recognized in field work and is described as 'seeing only what you are looking for'. Students will find that their field courses incorporate certain verbal, numerical and graphical techniques for minimizing this pitfall.

As a verbal exercise the student may be asked to assess the efficacy of a given classification for its purpose; this necessitates a directed search of the environment for departures from the pattern. Another and even more useful approach is the comparison of two classifications. Here two sets of concepts are considered *vis-à-vis* a single set of percepts, i.e., a single environment. For example, does it make better sense to classify soils by profile or by texture?

On the numerical side, counting or measurement may often reveal that impressions are misleading, and that the unaided eye tends to generalize a pattern with more consistency than it really possesses. This may be illustrated by reference to the shingle beach at Sandown, Kent, where flint pebbles appear to have been colour-sorted into distinct blue and orange bands. Counting reveals that the blue bands contain almost as many orange as blue pebbles and that the overall effect of blueness is due to the latter's generally larger size. In this case quantification is able to define an additional factor, size, that is not apparent upon first inspection.

Mapping, also, is a method of securing greater precision. A prevailing optical illusion in the study of erosion surfaces is their apparent extensive accordance of level when seen in profile. Committing them to a map usually reveals greater fragmentation and height variation, indicating more advanced and complex erosion than is apparent from visual observation alone.

In this second strand of field work the student will find a different balance between passive and active participation. Demonstration remains important and is the best means of learning precise criteria for established classes, but discovery, too, is gaining prominence. Its creative aspect is still played down because results cannot be guaranteed, but it now also possesses a critical aspect which enables the student to engage in purposeful appraisal of the demonstrated classifications.

The classificatory strand clearly involves more preparation than exploratory work and also a different kind of preparation. The great explorers went prepared with supplies, medicaments, and gifts to placate hostile natives. The classifier goes prepared with intellectual equipment: classifications or observational models. Field work is no longer a matter of simple preliminary fact finding. It integrates a part of the analysis in the observation and is thus more firmly embedded in the subject. This integration becomes even more marked in the remaining strands and accounts for the steady growth of field work as a feature of university geography.

Relationships

Although classification can vastly expand the mind's capacity for knowledge it, too, can reach a limit beyond which progress depends on finding some device for making patterns out of classes in the same way that classes make patterns out of units. This higher pattern-making device is *explanation* which can establish relationships between both classes and units, and is the basis of the third strand in field work.

Explanatory patterns tend to be clearest when they embody exact mathematical relationships, and some of the earliest work on geographical relationships belongs to what is now termed mathematical geography. One of the first recorded instances of geographical field work is the attempt made by Eratosthenes (276-196 BC) to establish the length of the earth's circumference by measuring the distance between Alexandria and Syene and comparing the sun's maximum noonday altitude at the two places. This result was remarkably accurate for his time.

Exact or fairly close mathematical relationships produce consistent or nearly consistent effects in the landscape with the result that explanatory patterns can be discerned from relatively limited amounts of information. Such patterns have strong predictive value; their extensions elsewhere can be readily postulated and then verified or disproved. Examples are furnished by the relationship of rainfall to desert, grassland and forest cover.

The less mathematically strict the relationship, the more difficult it is to observe and identify. It is no accident that the more complex and less precise and less predictable patterns of human cultures had to wait longer for explanatory analysis than did those of many physical features. Pattern making of this sort is more demanding. It needs far more field work to furnish a larger sample of the total pattern, and more profound insight to recognize its existence. The pattern makers had to rely less on numeracy and more on the power of words to crystallize concepts.

The great founding pattern maker of human geography was the German, Friedrich Ratzel (1844-1904). He developed the idea of mapping co-variants of human distribution, and also the twin themes of the influence of the environment upon man's activities and the influence of man in creating a sequence of cultural landscapes over time. He also stressed the need to measure man-land relationships. Each of Ratzel's themes subsequently developed into a separate school of thought known respectively as *geographical determinism* and *geographical possibilism.*

Determinism held the field at first; it was more able to model itself upon the causational explanations that had been evolved previously for more mathematically precise relationships. For example, climate was invoked as a main differentiating cause. Extremes of heat, cold, drought or deluge, and even extreme climatic monotony were suggested as inhibiting cultural and economic development, while temperate conditions permitted advancement and variability actively stimulated it. Consequently, although the level of human civilisation was a far more complex variable than vegetation, it was believed to have a similarly logical distribution in response to climatic conditions.

The deterministic hypothesis was far less simple than this brief outline suggests, but it had a profoundly simplifying effect upon the organization of geography's subject matter and allowed a vastly increased range of field observation to be related meaningfully. Other types of inhibiting or stimulating factor, such as soil quality, ruggedness of terrain, or accessibility, were also seen as determining human response, and since these differentiated much smaller areas than climate, they encouraged geographical interest in the homeland as well as in exotic countries.

The focusing of field-work attention upon home as well as foreign geography allowed the deterministic principle to be examined much more closely and critically, so that instances of departures from it began to multiply. Cases where man responded in the same way to different environments or in different ways to the same environment led the French, in particular, to reject the idea that man was controlled by Nature. In its place they pursued the alternative explanation, that man himself was the controlling agent and that he determined the landscape, creating something very different from that which nature had originally provided. The possibilistic model is associated with the name of Paul Vidal de la Blache, whose seminal work *Tableau de la Géographie de la France* (1903) inspired a classic series of regional field studies.

In Britain, with its traditional genius for compromise, deterministic and possibilistic approaches were both incorporated into field studies without any very acute sense of their mutual exclusiveness. Thus, a particular physical environment was discerned as influencing man's response in certain ways while at the same time the man-made landscape was analysed into component features and stages. Man's reaction to his environment and his action upon it was combined in a model of interaction, which ultimately became known as *geographical probabilism*. (O. H. K. Spate, 1952).

Probabilism can be thought of as a spectrum. At one end are those cases that are purely deterministic and at the other those that are purely possibilistic, while in between are the great majority which reflect the whole gamut of varying probabilities that arise from the differing degrees of influence exerted by numerous interacting factors.

Probabilism is a satisfying explanatory model in that it appears to be closer to reality than its predecessors, but it also poses severe practical difficulties when one tries to isolate the relevant influences and assign an accurate weighting to each one. Yet this has to be done if probabilism is to be more than a platitude. Consequently a great deal of the field work that the student encounters at university today is related to the resolution and control of multifactorial complexity. This is a task that again needs literate, numerate and graphicate approaches.

On the literate front there is a need to appreciate the theoretical advances made by social scientists, including human geographers. These can be summarized as the recognition of socially and economically determined patterns in man's actions, which are more susceptible to rational generalization than allowed for in possibilism. The crystallization of these ideas as explicit concepts enables the field geographer to isolate a new range of specific meaningful occurrences from what was formerly a vague web of complex interactions. Cultural attitudes may be taken as an example. Many societies reserve their most trusting attitudes for close contacts, variously the family, professional colleagues, etc., and show progressive caution or even hostility towards strangers, such as 'them' (the government), different religious groups or foreigners. In some cultures this hostility gradient is very gentle and the difference in attitude at each end is not marked. Other cultures exhibit rigid gradations with severe distrust of different castes or neighbouring villagers and overt hostility towards government advisers. Clearly the first type is likely to accept and profit by the diffusion of technical innovation, while the second type resists outside ideas and tends to remain in a more backward stage of economic development.

Once defined, similar contrasts can be recognized within quite short distances. For example a field group in Puerto Rico reported completely different attitudes in a slum of home-made shacks and a modern housing estate. In the former the inhabitants welcomed the investigators with smiles and spontaneously invited them into their homes, while in the latter they refused to communicate in any way. Differences of this kind can also be detected in British environments. The influence of the built environment in differentiating social attitudes has been explored by Jane Jacobs in *The Death and Life of Great American Cities* (1961) which is useful reading for any geographer attempting urban field work.

The verbal trend towards the better isolation of concepts is paralleled on the graphicate front. Field mapping is at pains to isolate sets of occurrences that belong to the same class, leaving other sets to be mapped separately. Although this approach is not new, it has become more dominant in recent years by expanding into situations where students were formally asked to observe relationships directly. For example, variations in land use were directly related to variations in physical terrain. This is now recognized as an approach that actively invites the pitfall mentioned earlier, the tendency of the eye to overgeneralize in its pattern making. Greater objectivity can be ensured if each distribution is collected and analysed as a separate entity in its own right before any attempt is made at correlation. Correlation is now regarded as a conclusion to be reached through analysis and not as a matter for direct observation.

As well as being isolated by type, each mapped occurrence should also show its location, date and size or quantity. Location is the essential basis for distribution analysis, and if it is not possible to include the full number of occurrences, then care must be taken to use a field sampling technique that will not introduce any distortion, or bias, into the distributional pattern that emerges on the map. The date requirement may mean no more than the inclusion of all the occurrences that exist at the time of the survey, but for man-made structures historical dating is often essential. For example, a map of hydroelectric power stations in the Alps ought to differentiate the occurrences by date in order to reveal the changes in site and size that have developed over time.

The size or quantity requirement links the graphicate to the numerate approach. The latter has become extremely important in present-day field work for two reasons. The first is its precision, which checks the over-generalizing tendency of the eye and hence increases objectivity. The second is its capacity, which permits multifactor analysis that is beyond the scope of the eye altogether. Appropriately enough it is probability mathematics that is employed to analyse the data related to geographical probabilism, and modern field work is strongly oriented towards obtaining the type of objective quantitative data that probability mathematics —statistics—can handle.

At present quantitative field work is in its infancy and is more at home with simple point occurrences that can be counted than with line and area features that have to be linearly or areally measured. Pedestrian and traffic counts figure prominently in student field exercises and prove to be relevant to a wide range of situations. A recent example is the counting of people and parked cars to investigate levels of recreational demand within Dartmoor National Park. This, however, was accompanied by another trend, the substitution of air-photo interpretation for field work when the photographs are sufficiently explicit and can save time, effort and expense. They may also improve accuracy by capturing one instant of time, as opposed to field work where the situation may change before the count has been completed. On the other hand, field work remains essential where it is linked to sample interviews, e.g., when every fifth car is stopped to enquire its origin and destination.

Current advances in the spatial analysis of lines and shapes suggest that these, like points, will figure more largely in the field work of the future. Some will be measured in the field; others will probably be mapped for subsequent measurement in the laboratory. Area, for example, is extremely difficult to measure in the field, but can now be derived from maps, much more easily. This emphasizes that the role of field work is not fixed, but can be adapted to supplant or be supplemented by other techniques as geography develops.

In correlating two or more field distributions, statistical work can reveal how far one distribution is explainable in terms of another or several others, and also how far it is left unexplained. This latter property is of the utmost value because it draws attention to the existence of further factors which, although still unknown, are not now overlooked. The indication of their existence can be used as feedback to stimulate the exploratory strand in renewed basic searching.

Statistical 'explanation' does not have the same meaning as explanation in every-day life. It establishes a relationship or association between two sets of occurrences, or variables, but it does not state whether this relationship is causational. It may well be nothing more than spatial propinquity, temporal simultaneity, or sheer accident. Even if a causational link exists, it is not stated which variable is cause and which is effect, or whether both are independently related to the same external cause. In order to establish causation the field worker has to progress beyond relationships *per se* to a higher order of pattern making, the study of systems.

Systems

The most significant feature of a system is that its parts are causationally related to each other and form a *functioning whole*. For some purposes it is appropriate to treat the whole as if it were a unit; for other purposes it is necessary to dissect out its component parts in order to establish how it operates and what the causational links actually are.

The first major system to be recognized in geomorphology was the Davisian cycle of erosion which replaced the relationships approach of earlier workers such as G. K. Gilbert by systematizing erosional processes in the spatial context of structure and the temporal context of stage. The discovery of a new system often exerts intense intellectual appeal, and the Davisian cycle was no exception. Many living geographers were first attracted into geography by the fascination of this particular piece of dynamic pattern-making.

The concept, or model, of a system does not spring into existence fully fashioned but begins as an outline offering much scope for field work. On the constructive side there is a need to complete the system's structure by investigating its dynamic links—processes, flows and feedback mechanisms—while on the critical side there is the need to modify it in the light of observed exceptions and departures. Davis himself worked assiduously to improve and modify his cycle, setting up parallel versions of it to satisfy conditions where its components assumed differential dominance. Thus, arid, glacial, coastal and karstic cycles were postulated to fit the respective dominance of drought, cold, waves and subterranean water.

The Davisian cycle inspired a great many other workers. In Britain it was a major factor in the growth of field studies as an integral part of undergraduate geography. In many countries it stimulated research into modifications which were gradually structured into independent rival systems: W. Penck's morphological system, featuring a dominant role for the forces of earth movement; L. C. King's pediplanation cycle, featuring backwearing as more important than downwearing; the system of climatic geomorphology developed by French and German workers; and the morphometric system of N. Strahler and others which stresses the importance of three-dimensional geometrical and other mathematical relationships. These are all of great advantage to the study of landforms because they can be compared critically in the field.

Today's students are particularly concerned with the field investigation of processes. Davis approached this task by means of comparative statics; he compared large numbers of existing landforms and arranged them in order of postulated development to show how initial surfaces can gradually change into something very different. This is now considered inadequate, and it is no longer assumed that a landform changes in a particular way unless the actual processes responsible can be established both qualitatively and quantitatively. Present-day research concerns itself with materials and the way they move in relation to fixed surface stakes, or deeply sunk inclinometers; the way they are sorted and arranged by dynamic agencies such as waves, rivers and ice; and the physical properties such as sheer strength which help to determine such movements and arrangements. Not unnaturally much active work is concentrated into situations where processes are rapid enough to measure over relatively short periods.

Graphicacy, literacy and numeracy all have a role to play in the systems strand of field work, as in the earlier strands. Much graphicate effort is devoted to systems analysis in the form of diagrams or flow charts which attempt to identify all the components and links, and all the inputs and outputs of energy and materials, as a basis for field investigation of those that are imperfectly understood. On the verbal front logic is a necessary tool to structure thinking about alternative propositions, and its mathematical equivalent, symbolic logic, is important for ensuring that relationships are explained in terms of function or causation, and not in terms of association only.

Systems are more characteristic of systematic geography than regional, and even within systematic geography they are more characteristic of physical than human branches. This is related to the nature of change, of which there are two kinds. Change in the form of process is an inherent part of a system; it does not detract from its unity and stability. The system may develop but its nature remains the same. The system may decay but a succession of similar ones replace it as in the case of weather systems.

However, change is not necessarily of this orderly nature. It may be unstable and disruptive, destroying the system without necessarily introducing another in its place. This is not process but transformation, a more complex level of mental product, which constitutes the fifth strand of field work.

Transformations

Field studies of transformations are concerned to be *diag-*

nostic of trends that disrupt rather than support existing systems. Frequently, although not invariably, such transformations are the work of society rather than of nature.

Man's intervention in physical systems is limited by the puny scale of the energy that he can exert in comparison with that exerted by nature. The answer to the myth that atomic explosions have produced climatic aberrations is the simple fact that the largest hydrogen bomb releases only a fraction of the energy contained in a single meteorological depression. The bomb's energy is quickly dissipated; it does not survive long enough to transmit repercussions through all the causational links of the atmospheric system as a true transformation would have to do. It is mainly those systems that are already in a state of precarious balance that are susceptible to transformation by human interference. For example, the pre-Cook ecosystem of New Zealand was only delicately preserved with the help of isolation; today it is estimated that 35 per cent of the country's vegetation consists of introduced species. Similarly, large areas of North America's high plains could only maintain and not recreate a grass cover; once ploughed up they rapidly degenerated into desert. These two ecological examples indicate that transformations may be either positive and trouble-free or negative and problem-ridden.

Transformations observed in the field of human geography may also be positive or negative; the latter have become more frequent with the more rapid and intense changes of the twentieth century. Civilizations and cultures are tremendously complex systems which are as yet imperfectly understood. Consequently many changes introduced in the name of progress have had unpredicted repercussions through the system's linkages and have sometimes done more harm than good. For example, in some cases where material advancement has been introduced without the moral advancement needed to support it, the result has been the decay of mores leading to unrest, violence and crime.

The geographer's concern with transformations, whether positive and progressive or negative and regressive, is with their manifestations in the spatial organization of areas. Field work on this topic began as serial surveys that showed conditions at intervals of time, but techniques have now been developed for synoptic diagnosis of change. The graphicate approach is basic for the study of problems such as blight in cities, land-use conflicts in the urban fringe, the extension of agricultural frontiers in developing countries or the toll taken by soil erosion in the wake of over-grazing.

The verbal approach is essential for elucidating the reasons behind the trends by interviewing the people who contribute to them. The field geographer must be conversant with questionnaire techniques and skilled in preparing questions that will avoid influencing the interviewee's response. He must also incorporate checks on veracity and other safeguards.

Both map and questionnaire are bases for the numerate approach. Their findings should be quantifiable and capable of statistical analysis so that trends my be diagnosed objectively and reported on with accuracy.

Implications

From the diagnosis of problems it is but a short step to the *prescription* of solutions and the last strand in field work is a part of applied geography, to be discussed in a later chapter. Application is based on implication. This is the name given by the psychologist to the highest of the six orders of mental product that have been used as headings in this chapter, and it is equally appropriate for the sixth and highest order of geographical pattern-making.

Planning problems are caused by negative transformations, usually in one of three contexts. Frequently the transformation has disrupted a previously stable system, such as a viable economic system, a well-knit social system, or in the case of a conservation problem, an ecosystem. Sometimes it may have replaced a pre-existing positive transformation, as when competition elsewhere precipitates industrial decline in place of steady economic growth. And third, a problem-ridden transformation may co-exist with either a stable system or a trouble-free transformation, as when derelict land is a by-product of some prosperous economic activity.

What the planner proposes as a cure is always another transformation but unless the implications of this have been thoroughly thought through the result may not be what is intended. It may seem wholly beneficial to transfer a slum population to a new estate of tower blocks, but in practice the benefit has frequently been vitiated by a weakening of social structure and a growth of juvenile delinquency in the new environment. In other cases the problem is solved locally but shifted elsewhere, as for example when urban renewal in the centre of a city creates a surrounding ring of blight.

Ideally the proposed planning action should establish a new stable system or positive transformation but this means a better understanding of systems and transformations than has been achieved hitherto. And because systems are built on relationships, relationships on classes, and classes on unit facts, planning studies call into play the whole range of field activity.

The highest strand is needed to establish the implications of proposed transformations; this involves studying the results of earlier planning action in as many different field environments as possible in order to strengthen cost-benefit analysis and decision making. There is also a need for the remaining strands down to the lowest. Planners need the open-mindedness inherent in straightforward fact-finding surveys because this may afford a first clue to the missing links in the system to be created. In field work the six strands are not separate but intimately twined around each other to produce a rope, the strength of which is its unity.

FURTHER READING

CHORLEY, Richard J. and HAGGETT, Peter (eds.), *Frontiers in Geographical Teaching* (Methuen, 1965).

STEERS, J. A. (ed.), *Field Studies in the British Isles* (Nelson, 1964).

6

Hypotheses and Models

Cuchlaine A. M. King

In many universities, geography can be studied in three faculties: arts, science and social science. This reflects the fact that geography functions as a bridge subject between different aspects of our culture, both in its content and in the methods of approach that are being developed to study it. However, the methodology is increasingly biased towards science since it is the scientific method that seems capable of achieving the most notable advances.

Geographical subject matter, the earth's surface and the distribution of the many attributes that go to make up its characteristic landscapes, covers a wide field. It ranges from rocks and landforms, through vegetation, climate and soil to man's use of these and other things. It is not a static study, however; movements, growth and change are characteristic of most geographical elements, and a study of the dynamic aspects of geography has given rise to many new techniques. Geography is therefore an active and dynamic subject that can best be studied scientifically by means of hypotheses and models. This chapter aims to give some indication of the many new and exciting methods that are being developed in this field.

The scientific method can be briefly summarized as five steps:

(a) An objective or hypothesis must be formulated.

(b) Data relevant to the hypothesis must be collected on the basis of an observational model.

(c) The data must then be prepared and stored in a suitable form, such as on a map, in tables, or in a computer.

(d) The data must be analysed or processed.

(e) The results of the processing must be used to reach a conclusion concerning the original hypothesis. This may lead to the formulation of a law or the establishment of a fact depending on whether the data are general or particular in character.

In the past geography has often dealt with the unique rather than the general. If a particular area, unlike any other, is studied, hypotheses concerning it can lead only to facts or relationships that apply to that area alone, and no general laws can be formulated. More recent studies have often concentrated on the patterns and relationships that do have a wider application. Such patterns and relationships can lead to theories and laws.

Unique areas or topics are likely to be large, complex and varied, and lend themselves only to deductive study by theoretical argument. Much more useful studies can be made by dealing with smaller units that occur in greater numbers and that are less variable in character. From a large number of cases or individuals valid generalizations can be made by inductive reasoning. For example there are many rivers or villages, so that useful laws can and have been developed concerning them. Each river or village is itself unique, but it has enough in common with other rivers or villages to allow it to be grouped with them for purposes of study. But there is also a considerable range of variability amongst rivers or villages, and this variability gives much scope for classification. A great many sophisticated methods of classifying the raw material of geography have been developed, and are being increasingly applied.

Hypotheses

The formulation of a hypothesis presupposes a problem to be solved. The hypothesis also needs to be capable of testing if it is to yield a worthwhile conclusion. There is a danger in working with only one hypothesis that should not be overlooked. The hypothesis may suggest a certain relationship, and in seeking data to test its correctness it is easy to leave out those cases that do not support it. For this reason the method of multiple hypotheses has many advantages. As many hypotheses as possible are considered, and the one that best agrees with the data collected becomes the working hypothesis or theory. An example of this method of approach is the work done to elucidate the processes that give rise to the intriguing patterns that develop in perennially frozen ground. As early as 1927 no less than 25 different hypotheses had been proposed to account for the polygonal features characteristic of permafrost areas. It was not until much data had been collected by prolonged field observations, by careful laboratory experiment and theoretical reasoning supported by mathematical calculations, that the features could be assigned to different classes that have been formed by different processes. Some of them have been shown to be the result of extreme low temperatures causing contraction and cracking, while others are formed by the stirring-up action of frost and ground ice. These studies illustrate the value of careful scientific observations carried out in many different ways, by use of a wide variety of techniques to test the validity of the many conflicting hypotheses. This example was drawn from the physical landscape, but problems in human geography can be treated by the same scientific method using very similar techniques and approaches. Both major branches of the subject have much to learn from each other's approaches and methods.

Data collection is essential in most studies; although some may be purely theoretical, even these must be based on facts derived from earlier data collection. The value of dealing inductively with hypotheses has been mentioned, and the resulting generalizations are increasingly valid as they are based on more and more cases. However, there is a limit to the number of cases that can reasonably be dealt with. Too many cases lead to unwieldy data, and the work involved is not justified by the greater certainty of the results. On the other hand, if there are not enough cases the results are not valid. The problem, therefore, is to select the optimum number of cases from all those that are available. This is a matter of sampling design and falls within the field of statistics. A small preliminary sample can give information that allows a specific sample size to be calculated at any required level of confidence. The level of confidence indicates the percentage of times the results would be expected to occur by chance. The size of the sample required depends on the variability of the objects sampled. If they are all exactly the same, then one is all that need be measured, but the more they vary the greater the number that is needed in the sample. A great deal of time is saved and much more valid results usually emerge if the sampling plan is considered very carefully when the hypothesis is first formulated. The nature of the hypothesis determines the nature of the data sampled. Thus if one is studying the effectiveness of glaciers to erode stones, one must take stone samples in different positions relative to the glaciers; or if the study is concerned with town functions, different types of towns must be sampled.

When the data have been collected they must be prepared for storage until required for processing. Storage must be in a form suitable for subsequent retrieval. In the study of slopes, for example, the data may be collected by surveying a sample of slopes in the field and collecting samples of soil from the surveyed slopes. The figures in the field notebook must be converted into heights and distances from which profiles can be constructed and angles of slope calculated. The soil samples must be analysed in a laboratory and the results converted into meaningful values, such as the average size of material and the spread of sizes in the sample, i.e., its degree of sorting. These results must then be associated with the profiles to which they refer. Such data can often be conveniently stored in matrix or table form, or on punched cards or similar devices. In some instances the data may be stored in map form, e.g., the function of buildings in a town can be recorded cartographically by the use of suitable symbols. A very elaborate method of data storage is being developed by the Land Inventory of Canada. The data are originally recorded on maps, which show the distribution of rock type, soil type, relief, land use, land potential and other categories. They are prepared for computer storage partly by the use of a drum scanner which records the boundaries of delimited areas, partly by the use of a digitizer which assigns a location to each area, and partly by punching code numbers on cards to record the categories of variables that the areas represent. When required, the class of each variable can be retrieved for any point on the map that is specified to the computer in which the data are stored.

The foregoing emphasizes an important fact concerning geographical data. This is its multivariate character. Each part of the earth's surface is characterized by a great many variables. Each feature, whether it be a coastal spit or a factory, is influenced in its growth and development by a great many different processes, the action of which may change over time. It is this multivariate nature of geographical data that makes the subject both difficult and interesting. It is this characteristic in the data that necessitates both careful formulation of a hypothesis or preferably several hypotheses, and a well-thought-out sampling plan.

Data processing in geographical analysis is best carried out by quantitative techniques because these usually lead to more useful results that can be expressed in statistical terms with a definite level of confidence. Thus a relationship between two variables may be stated in terms of a 95- or a 99-out-of-100 chance of its being a genuine relationship. When data are analysed quantitatively the calculations may be long and tedious if there are many members of each sample. Consequently the computer is being used increasingly to do the relatively simple but time-consuming calculations that used to be done on a slide rule or desk calculator, or even with paper and pencil. In the analysis of multivariate problems the computer is essential as even the simplest situation with only about four variables would take far too long to analyse by hand. Because it can process multivariate problems so rapidly, the computer has led to a great increase in the sophistication of geographical studies, thus making them more rewarding and more scientific in their methods and results. There are many statistical and mathematical methods that can be used for different types of data analysis. They all provide results that indicate the interrelationships between the data that are being processed. These interrelationships either support the hypothesis being tested or show it to be inadequate or unacceptable. If the results confirm the expectations of the hypothesis, then the problem may be explored further by collecting more data to deepen the understanding of the situation. If the hypothesis is shown to be unacceptable, then another one must be developed. In this way the scientific method is applied to geographical studies and, from the successful hypotheses, theories and laws may be developed. The law increases in generality as it applies to wider and wider situations. In human geography in particular the situation is so complex and the number of variables so large that laws are rather few in number, although some have been suggested. These must always be stated in probabilistic terms because the many variables that are involved are not amenable to control, nor can the conditions to which the law applies be stated theoretically, as is possible with many of the laws of physics. Nevertheless the relationships that are established can be used to predict future situations.

The ability to predict is an important property of a successful hypothesis or theory, as it allows the results of geographical study to be of practical value in solving real problems. A knowledge of the causes of coastal erosion, for example, enables successful measures to be taken to combat erosion without resulting in damage to the coast elsewhere. It is in such situations that the geographers' appreciation of the spatial aspect of the problem is of value. There are also many situations in which economic, social and urban geographers can use the scientific method to predict the likely result of a particular course and hence aid planning for the future.

Models

The great complexity of geographical situations has been stressed. It is partly this quality that has led to the adoption of models in geographical study. This word, which is used with a different meaning as a noun, verb and adjective in ordinary everyday language, is also used to cover a very wide range of different ways of studying geography. Models pertaining to different branches of the subject have been discussed in detail in the book *Models in Geography*, edited by R. J. Chorley and P. Haggett. The aim of all models is to simplify a complex situation and thus render it more amenable to investigation. The complexity of geographical situations is such that models are of particular value in studying geography. Some examples of the different types of models that are being used in different branches of geography will be considered. They can be discussed in an order of increasing abstraction. (a) scale models, (b) maps, (c) simulation and stochastic models, (d) mathematical models, (e) analogue models, (f) theoretical models.

Scale models, also called hardware models, are perhaps the easiest type to appreciate as they are direct reproductions, usually on a smaller scale, of reality. Hardware models may be either static, like the model of a land surface of a geological model, or dynamic, such as a wave tank or river flume. Dynamic models are perhaps the more interesting and useful in geographical work. The great advantage that a model of this type has over reality is that the operative processes can be controlled. This allows each variable to be studied separately. In a wave tank the effect of material size, wave length and wave steepness on a beach slope can be measured quite accurately if two variables are held constant while the third is varied. If the resultant beach slope angle is plotted against each variable in turn the points obtained in each case may either fall in a nearly straight line indicating a significant relationship, or in a diffuse scatter suggesting little or no relationship. Close relationships revealed by the model may not be apparent on a natural beach where the wave variables cannot be controlled.

There are, however, difficulties in applying the results of model studies of this type to the natural situation. One of these is the problem of scale. If wave size and material size are scaled up in the same proportion then the sand of the model would become large cobbles in nature—and these two materials do not react similarly to waves. Again if sand in nature is scaled down to model size it would be silt or clay, which also responds differently from sand under wave action. Despite such difficulties scale models have yielded very useful results in many fields of enquiry. The fact that engineers nearly always make a scale model before embarking on any major project such as a river improvement or harbour works scheme, demonstrates the value of this type of model.

Geomorphologists have carried out fundamental research with models in order to investigate processes that are difficult to observe under natural conditions, such as marine processes and river action. An interesting new type of scale model is a glacier made out of bouncing putty that closely simulates the Malaspina glacier moraines when the putty is made to move by whirling it in a centrifuge. The initiation and movement of turbidity currents that are important in the study of submarine canyons have been reproduced in models and these have given rise to new theories concerning the processes. On a larger scale, models have been developed to imitate the formation of whole mountain ranges by convection. Another ingenious device has been used to study the movement of the North American Pleistocene ice sheet. This consists of an electrical conducting sheet, and the investigation is based on the relationship established in physics that flow equals driving force times acceptance. The area studied is drawn on a special type of paper, Teledeltos, on which the four sources of ice and the marginal 'sink' are marked with special paint. When an electric field is switched on it is possible to draw in equipotential lines between the sources and the sink. The streamlines of the ice are at right angles to these equipotential lines. This model shows very good agreement with what is known about the actual direction of ice flow. It seems possible that a similar type of hardware model could be used in the field of human geography to study population potential.

Maps are the 'models' that are most familiar to geographers. They are a special type of scale model which becomes increasingly abstract as the scale becomes smaller. At one end of the spectrum is the stereo-pair vertical air photograph which provides virtually a true scale model of the real world. It is, however, static and represents only the area shown at one instant in time. A single vertical air photograph loses the impression of height but still shows all the visible elements of the landscape virtually true to scale. A large-scale map loses much of the detail of the landscape although it can show buildings, roads and other features of this size accurately. As the scale is reduced the information becomes more symbolic and can no longer be shown true to scale; ever more detail must be omitted. The map can, however, give an indication of the relief by means of contours, hill shading and hachures; this is missing from the simple vertical air photograph. The map also has another advantage over reality in that it can show a very large area simultaneously, so that the mutual space relationships can be much more easily appreciated than on the ground. Many maps use symbols to show specific features or distributions, such as population density; these are even more abstract and further removed from the reality that they are representing. A still further abstraction is the loss of correct Euclidean space relationships that occurs in topological maps, such as diagrams of stations on the London Underground. They maintain their correct order, but distance and direction are wrong. A new insight into a familiar area can be given by drawing a diagrammatic map where the scale is not correct for area, but is adjusted to show the population or some variable to scale. Modifications in area, distance and direction are also needed in maps covering the whole world or large parts of it, as a curved surface cannot be correctly reproduced on a flat piece of paper.

Simulation and stochastic models have been developed to deal with dynamic situations rather than the static state shown on a map. This type of model simulates some particular process by means of random choices, hence the term 'stochastic', which is connected with chance occurrences. It can be illustrated by its application to drainage development. Starting with a pattern of grid squares it is assumed that a stream source exists at the centres of certain randomly chosen squares. Random numbers are again used to determine in which of four possible directions each stream will flow and a line is drawn to represent its course as far as the centre of the adjacent square. By repeating the process (with certain reservations that approximate to reality) there emerges a complete drainage network which shows many similarities to natural drainage patterns. The conclusion can be reached that the natural drainage pattern has some element of chance about its make-up.

Simulation models can also be of use as a means of analysing a large number of variables, a recurring problem in geography. For instance, the development of a coastal spit can be shown to depend on a number of distinct processes or wave types. These different processes can be built into a model in such a way that they are each allocated a specific range of random numbers. Each random number that comes up results in the operation of the appropriate process. In this way the spit can be built up by the action of the different processes in a random order, but in specific proportions. If the simulated spit resembles the real one, then one can conclude that the processes probably operate in the proportions specified in the model. Once a realistic model has been found it can then be used to predict future development of the spit, provided that the processes continue to operate in similar proportions.

Stochastic simulation models have also been successfully used in the field of human geography, to study the spatial diffusion of a variety of phenomena, including the spread of populations, diseases such as foot and mouth in cattle, or innovations such as the use of a particular piece of machinery. The simulation is made realistic by imposing barriers that can be crossed with varying degrees of difficulty. Random numbers are used to determine the direction of spread and the effect of the barriers can then be assessed. The term 'Monte Carlo' is used to describe some stochastic models, in which chance alone determines the outcome of each move within the conditions of the model. The Monte Carlo model may be compared with the Markov chain model in which each move is partially determined by the previous move. The Markov chain is exemplified in the random-walk drainage development model described above. Both types have been applied in many fields of geographical research.

Mathematical models represent the operation of specific

processes by means of mathematical equations which relate the operative process to the resultant situation. It is necessary, however, to have a sound knowledge of the physical processes concerned, and consequently this type of model building has been mainly the work of physicists. For example, a dynamic mathematical model of glacier flow has been constructed by J. F. Nye. He simplifies the basic assumptions as far as possible to make the equations sufficiently simple to solve. Thus the glacier bed is assumed to have a rectangular cross profile of uniform size and specific roughness. The ice is assumed to be perfectly plastic in its response to stresses. Then, given certain stresses, the response of the ice can be calculated by means of differential equations. These can predict specific flow patterns and ice profiles for given values of the assumed conditions. The geomorphologist can play his part by measuring the flow patterns and glacier dimensions in the field. The closeness with which these approximate to the calculated values is a measure of the success of the mathematical model. If the observed flow pattern agrees closely with the predicted one, then the model can be used with some confidence to provide values for flow in parts of the glacier that cannot readily be measured in the field, but which are very important in studying the effects of glaciers on the landscape. The speed of basal flow is important in this context.

Mathematical models have also advanced our knowledge of how rivers move their load and adjust their beds, and how waves operate on the coast. These models are usually in the form of differential equations largely based on known physical relationships, and it is essential to test their numerical results against observations made under natural conditions or in a scale hardware model. The models are only as successful as the assumptions and simplifications on which they are based are true and valid. They provide a very simplified situation, but one that can be expressed in precise numerical terms and hence is capable of suitable mathematical manipulation. For this reason such models are more suited to problems in physical geography.

There have, however, been somewhat different developments of mathematical models in human geography. These are more in the nature of empirical relationships that can be expressed in mathematical terms. An example is the rank size relationship. This relationship shows that within any class of occurrences there are usually a few very large items and many small ones with a fairly regular distribution between. It has been applied to towns in many parts of the world. There are a few very large towns but many more small ones, and between the two a moderate number of medium-sized ones; the relationship is approximately linear on a double logarithmic scale.

Mathematical models have also been developed in economic geography, which is more susceptible to quantitative formulation than other branches of human geography. Such models are often not dynamic in the same way as are the differential equations in physical geography, although some do deal with flow of goods, etc., from one region to another.

Another mathematical model is linear programming, which is relevant to many situations in economic geography. It is a method of finding the optimum solution to a problem in which several conditions must be fulfilled. A factory will have certain requirements of labour, raw materials, transport and access to markets, each of which determines conditions that can be expressed as mathematical equations and represented graphically as straight lines. When all the equations have been plotted they reveal the point of optimum value in terms of location. The procedure provides a definite solution based on the values assigned to the equations. If the values are accurate then the optimum solution will be provided.

Analogue models differ from those types of model that have already been described in that, instead of using imitations of the original or symbols to represent it, the feature being studied is compared with some completely different feature by means of an analogy. An analogue model uses a better-known situation or process to study a less-well-known one. Its value depends on the research worker's ability to recognize the elements common to two situations. These elements constitute the positive analogy; the dissimilar or negative analogy and the irrelevant or neutral analogy are ignored.

Reasoning from analogy has long been a part of geographical study. James Hutton in his major work published in 1795 recognized the similarity between the circulation of the blood in the body and the circulation of matter in the growth and decay of landscapes. A similar circulation can also be seen in the hydrological cycle. Great advances in understanding the development of certain geomorphological features have been achieved by the recognition of analogies. The work of W. V. Lewis on the formation of corries was based on an appreciation of the similarities between rotational shear slumps and corrie glaciers. The shear slumps take place on a cliff face or steep slope where weak strata underlie stronger rocks. The wetting of the weak beds allows the stresses to build up until they exceed the strength of the rock which gives way suddenly along a curved shear plane. The hard rock becomes tilted from its original horizontal position to dip back into the hillside. This feature, which resembles the up-glacier dip of stratification planes in the ice of cirque glaciers, suggested to Lewis that rotation may also be taking place in the glacier movement. Detailed observations of movement in a cirque glacier confirmed that the ice was in fact rotating about an almost circular arc. This insight, gained by perceiving an analogy and supported by a very accurate field investigation, allowed a hypothesis of glacier movement to be set up, confirmed and further elaborated.

The analogy used to further geographical knowledge must be better understood than the feature being investigated. The behaviour of metals under stress has been intensively studied, and this has allowed useful analogies to be drawn between metals and ice. Methods of dealing with one problem can often be transferred by analogy to a completely different situation. The study of kinematic waves has been applied to the movement of vehicles on crowded roads, the movement of stones and flood waves in rivers, and the formation of surges at a glacier snout. These very dissimilar problems have in common the fact that they are one-dimensional flow phenomena and from this point of view they can be treated by the same technique.

Analogies have also proved fruitful in the study of problems in human geography, for example, those that draw on certain well established relationships in physics. The gravity model is a good example of this type. It is based on the physical observation that the attractive force between two bodies is proportional to the product of their masses divided by the square of the distance between them. The value for the distance in the model is often squared to approximate more closely to the force of gravity as observed in physics. The attractive force may be considered in terms of transactions between two places. The number of transactions will be likely to increase as the size of the places, often measured in terms of population number, increases and as the distance between them decreases. This model presupposes that there is no other force involved to limit the transactions, such as an international or language barrier. Various other physical relationships used as analogue models include the patterns of a magnetic field and the second law of thermodynamics.

Theoretical models are the last type to be discussed. These can be subdivided into two categories. The conceptual model provides a theoretical view of a particular problem allowing deductions from the theory to be matched against the real situation. This can be exemplified by the theoretical consideration of the effect of a rising and a falling sea level upon the coastal zone if certain specific conditions are fulfilled. It is assumed that wave erosion is the only process operating, that waves can only erode rock to a certain depth, of the order of 40 feet, and that the waves erode a wave-cut platform to a certain gradient below which they cannot operate effectively. It is also assumed that the initial coastal slope is steeper than this gradient. A consideration of the prolonged action of waves eroding under these conditions, with a rising and falling sea level, leads to the conclusion that only with a slowly rising sea level can a wave-cut platform of great width be produced. The theoretical forms of the coastal zone under the various conditions specified can be established and then compared with actual coastal zones. Much more elaborate theoretical models of this conceptual type have been developed in the study of the evolution of slope profiles. These are based on the known or assumed effect of different slope processes. A long series of stages of modification can be derived from this type of theoretical model, and these can again be matched with actual slopes.

The second type of theoretical model is associated with the word 'theory', when this is used to denote the overall framework of a whole discipline. The framework must not be too rigid or it will cramp the growing edges of the subject, where the most exciting work is going on. The ideal is a flexible framework that can contain the wide variety of geographical endeavour and yet give it coherence and purpose. Models are particularly valuable in this context as they are often common to all the different branches of the subject and so help to give it unity.

An analogy may help to illustrate the way in which the vast and growing amount of geographical data may be organized within a theoretical framework. Geography may be compared with a five-storeyed building, each storey being supported by the one below and supporting the one above. The lowest storey is the one which accommodates the data, the raw material of geographical study. The data lead up to the level of models, where they are organized in a suitable way for analysis. The techniques of analysis, lying on the next storey, depend on the model adopted for the study. Analysis leads up to the next floor, concerned with the development of theories and the theories in turn lead up to the formulation of tendencies and laws. These are located at the top as they are the ultimate aim of geographical methodology.

FURTHER READING

CHORLEY, RICHARD J. and HAGGETT, PETER (eds.), *Models in Geography* (Methuen, 1967).

COLE, J. P. and KING, C. A. M., *Quantitative Geography* (John Wiley, 1968).

7

Systematic Geography

W. Gordon East

It becomes increasingly unusual for the professional geographer, whether as teacher, researcher or writer, to look the whole world squarely in the face; rather he elects to deal with some part of the earth's surface or with a single aspect of it. In the latter case, he is satisfied with studying one aspect only of a chosen area. Needless to say, these apparent retreats from the main task can be easily explained and justified. The surface of the earth, even though it constitutes a single, intercommunicating and interacting region —man's environment—and even though, with ever-improving means of transport and communication, it appears to grow smaller, nevertheless consists of highly complex, varied and interrelated phenomena: physical, biotic, and social. Not only does our world vary regionally and locally; it is also changing all the time, so that the geographer's skill and power of work can never hope to describe and account for it in any final or definitive way. The task itself is difficult, in view of the many special skills that it necessitates—but it is also challenging. Thus the geographer seeks to conquer his vast field by dividing it up and by engaging in those specialized studies which are variously known as 'systematic', 'general' or 'topical'.

The terms 'systematic', 'general' and 'topical', as applied to geography, need explanation. The first, which has become British usage, may well have been borrowed from the science of botany which used it to describe the work of classifying species of plants. It does not of course imply that other forms of geographical investigation, notably regional geography and historical geography, are conducted without either method or plan. H. J. Mackinder hoped that 'the expression systematic geography' would not take root, but this it has certainly done (*Geographical Journal,* Vol. 25, 1905, 312).

'General geography' is the oldest-established of the three terms and has a respectable pedigree; the term was used, explained and illustrated more than three hundred years ago by the German writer Bernhard Varenius in his *Geographia Generalis* (1650). In 'general geography', mathematical and scientific concepts and reasoning can be applied and the attempt is made to study single aspects of the earth's surface or its parts. Varenius was at pains to distinguish 'general' from 'special geography' where the geographer tries to describe a whole country and is necessarily much concerned with human societies. The third term, 'topical', has a certain currency among American geographers who use it for studies of specific topics.

The advance into scientific borderlands

It used to be said that, if all the other disciplines fully undertook and achieved their proper objectives, too little might be left to geography to justify for it a dignified place in the world of learning. But in fact the many other disciplines fall short of totalitarian effort; they leave many borderlands neglected or insufficiently developed, and into many of these geographers have profitably advanced. Thus it might appear that geographers undertake systematic studies which tend to extend and enrich not only geography itself but also a range of other disciplines with which they have forged close relationships. Let us explore these external relationships before considering in what ways, by what means and with what success, geographers engage in systematic studies to serve their own ends.

A randomly selected list of systematic topics might include the following: the spatial distribution of world population, the mediterranean type of climate, erosion surfaces in Wales, the industrialization of Britain, rural depopulation in the Scottish Highlands, landlocked states, seaside resorts, coastal landforms, rural settlement forms, and the origins of towns in Egypt. The reader will immediately note that these are single-topic studies, some of which relate to the whole world, others to a specified part. He will note too that the selected topics are by no means exclusively of geographical interest; indeed a few of them might appear to him only marginally geographical. In varying degree the topics listed engage the interest and special skills of other scientists and scholars. Thus, that special kind of statistician called a 'demographer', whose business is with vital statistics (those relating to mankind, not to those of the fashion model), clearly has much of importance to say, so far as available statistics exist and can be interpreted, about the population of the world and its parts. So also the origin of towns in prehistoric Egypt could be more a topic for archaeologists than for historical geographers. And the artist and the literary man, not to mention the biologist, have much to contribute about coasts which is wholly distinct from the geographer's findings about them when he wears the mantle of geomorphology.

If then the geographer ventures into borderlands where others already claim squatters' rights, what does he seek to do, and what is he equipped to do, that demonstrably advances knowledge, and how far are the products of his work geographical?

Let it be understood at once that systematic studies in geography, called sometimes 'adjectival geographies', involve specialization and thus call for geographers specially equipped. Those who undertake such work at the research level have more than the usual need to understand one or more disciplines other than their own. The geographer who would succeed within the field of meteorology must know something of physics; he is concerned with energy transfers at different levels of the atmosphere, the pressures, temperatures and movements of air masses, insolation, radiation and much else, all of which are subject to the laws of physics. And physicists would normally regard meteorology as part of their field. Similarly, different forms of vegetation, which clothe so much of the land surface, are made up of individual plants and associations of plants, the study of which, ecology, falls primarily within the science of botany. Thus, geographers investigating plant geography or specific vegetation distributions, such as those of temperate grasslands and tropical forests, need botanical knowledge. It is evident that the various systematic geographies cannot ignore the methods and findings of the many relevant disciplines that fall into categories conventionally designated as Science, Arts or Social Sciences. Economics has relevance to economic geography, sociology to social geography, geology to geomorphology, mathematics to cartography, and history to settlement geography; in each case of geographical specialism there is at least one cognate study on which the geographer must in varying measure depend, if his advance into a borderland and his work there are to succeed.

Of course, it is not the whole of other disciplines that the systematic geographer needs to make his own; rather his need relates to some part of the cognate discipline and to its methods and assured findings. Thus the geological principles of glaciology or of stratigraphy, may be of cardinal importance to a geomorphologist in a particular investigation, while he may be able to ignore other branches of geology, such as mineralogy, and palaeontology. So also the social geographer needs selected aspects of history and sociology, the better to carry out enquiries of primarily geographical interest.

It should be noted that, although each branch of systematic geography mainly depends on one other subject, further subjects can also give aid in lesser degree. Thus the political geographer, following a line of investigation, may profitably consult not only works of modern history but also those of political science and international law.

The purposes of systematic geography

The purpose of systematic studies undertaken by geographers is not explicitly to advance the cognate fields which they explore. Were this so, their action might well be resented. Similarly, on the other hand, cognate scientists cannot be expected to appreciate mere borrowings of knowledge that do not lead to original contributions. When, however, it is recognized that the purpose of geographical excursions into the borderlands falls between these two extremes and that they are necessary and useful journeys in search of answers to geographical questions, then geographers' efforts may be welcomed as wholly or largely complementary to those of other scientists. Borderlands, because they are liable to suffer neglect, offer promising areas for exploration; and, in any case, since the divisions between knowledge are arbitrary and very much matters of convenience, it is of little significance which science yields results of interest and importance. However, it would seem right and inevitable that geographers should engage in systematic studies in the attempt to answer geographical problems and thus contribute to the understanding of man's habitat.

There has been a strong tendency in recent decades to increase the range of systematic specialization. Thus one systematic branch, economic geography has been split up into several parts: agricultural geography, manufacturing geography, transportation, mineral production, the geography of resources and marketing geography. Such 'adjectival' geographies can provide fields of study so substantial and of such intellectual and even practical interest as to become sufficient in themselves. Some of those who study them may wish to specialize so completely as to ignore the wider interests of their parent subject, and perhaps also to hive off from geography and to establish an independent niche in the world of learning. The reader who examines the contents of a leading journal of geography and notes the wide range of topics discussed might wonder whether geographers have become pluralists and are following their bents without co-ordination of effort and without a clear view of their central objective.

Without going as far as plurality, geography has long been held to possess a dual character: physical and human. Some support is given to the alleged duality by the way in which syllabuses and textbooks refer to physical geography and human geography, and to physical landscape and cultural landscape. This topic has been vigorously debated in recent years particularly by Soviet geographers who face an acute intellectual difficulty in that one part of their field clearly falls within the orbit of the natural sciences and the other part within the humanities, and that the methods used in these areas of learning are different. This duality is traditional. It was recognized in the earliest geographical literature of classical antiquity and it was sharpened and hardened by methodological thought in Germany during the nineteenth century when geography as a discipline was taking shape. Certainly there are difficulties in the study of a subject which has been, and is, variously grouped with the natural sciences, social sciences and the humanities. But the difficulties have to be faced because they spring inescapably from the nature of the geographer's task which is to try to understand man's terrestrial habitat as it changes regionally and locally. And these habitats or areas are composite in character, combining physical, biotic and cultural elements. To ease and advance study, these elements are separately considered, but they do not exist separately in nature. It is convenient but surely fanciful to pretend that the physical landscape of England is, strictly speaking, available for observation. There is only a single landscape, which reflects the results of both physical and social forces. In other words, since it is generally agreed that geography tries to describe and account for the areal differentiation of the earth's surface, then this provides the prime and central objective, vastly difficult though it is to attain, given the scale and complexity of the task.

The question raised above can now be answered. What looked at first sight like a plurality of scientific operations, each following its own line, is now seen to be subservient to the central purpose of geography. This central purpose has long been expressed in shorthand as 'regional geography', which is discussed at length in the next chapter. Brief reference to regional geography is necessary here since this is closely relevant to an understanding of the aims and purposes of systematic geography.

Varenius long ago made it clear that geography consisted of two integrated parts, the second of which he called special geography, later to be more familiarly known as regional geography, the term which Mackinder was at pains to establish. There can be little doubt that for a variety of reasons geographical studies should be made of whole countries or of areas large and small. Such studies satisfy human curiosity, about, for example, one's own country or the part of it in which one lives. There is, beyond doubt, much of practical usefulness which geography can supply by presenting an integrated, or holistic, account and explanation of a country or area. This usefulness is recognized by business men, industrialists, agriculturalists, travellers, and governments. The case for such studies need not be laboured and, earlier in this century, the efforts of geography concentrated, in teaching and in writing, on regional geography. It is now clear that this concentration, useful though it was in stimulating interest in a study new to the university faculties, often led to somewhat unsatisfactory results. While the best of such efforts produced such classics of geographical literature as Mackinder's *Britain and the British Seas* and Vidal de la Blache's *Tableau de la Géographie de la France,* an excessive diversion of effort to studies of small areas, dignified as 'regions', was not too unfairly rated by an American geographer, V. C. Finch, as 'a monkish retreat upon minutiae'. In short, regional geography, the hardest geographical work to excel in, since it requires not only scientific knowledge but also sound judgement and literary art, was, in many cases, building an elaborate and complex structure on inadequate foundations—foundations which could be laid properly only in the light of further work in the systematic branches.

The duality between the geography of areas and the geography of specific topics related to areas is more apparent than real. Their essential unity is exemplified in the fact that any valid synthesis of the geographical features of any area so as to reveal its character and potential must rest on the established findings of a wide range of systematic studies. One justification for systematic researches is thus evident. By concentrating efforts on a particular aspect, be it soils, climate, landforms, settlement or agriculture, the geographer can improve methods and refine concepts, thus clarifying many features which, when interrelated in area, form the subject matter of regional geography. Although the systematic geographer is at times deeply involved in problems of methodology, and although he may appear absorbed in the problems of studies cognate to geography, he must necessarily study areas; his work is a partial areal geography. After all, the geographies of soils, landforms or climate, given that each can arouse and sustain curiosity and intellectual excitement, present only arid knowledge until they are related to the environments (physical, biotic and social) in which we move, of which we are part, and which form the object of our study. It is right, then, to conceive of the two parts of geography as two necessary approaches, without which the discipline atrophies.

Systematic geographies have some measure of importance as applied sciences, contributing towards the understanding and solution of current problems or planned developments in a variety of fields—economic, social and political. Here it might appear that systematic geographies make their own independent way. Consider two examples. A geomorphologist with sufficient local knowledge could indicate the location, extent and character of gravel deposits within a certain range of London, and such expert knowledge has a practical, indeed a business value. Similarly, a chain store, wishing to expand its activities in a given country, could not do better than consult an urban geographer with knowledge of that country's towns, to find out in which it might set up new stores, and also what size these stores should be. But notice that, in this applied field, as distinct from the pure field where knowledge is sought for its own sake, systematic geography is concerned not only with its own specialism, but also with knowledge of particular areas. In other words, the duality of geography disappears again and systematic geography finds itself necessarily linked with regional geography.

Two systematic examples

This discussion may now be illustrated by brief reference to two systematic geographies which appear to lie widely apart yet are both integral parts of geography. The two selected are geomorphology, which has been defined as the science or genetic study of landforms, and political geography, which is concerned with the relations between geography and politics. The first clearly has much to do with physical processes, the second with human activities. The first belongs to the natural sciences, the second to the social sciences. Both have a common purpose in that they are trying to illuminate single aspects of the world in which we live; they are also similar in that they have to look back to past events and processes in order to understand what now exists. The attempt is now made to indicate the objectives of each, the methods which they employ, and the success achieved.

Geomorphology in Britain falls mainly to geographers, although it began as a part of geology and some geologists, notably in the United States, are no less engaged in it. Although the landforms of any part of the earth are in fact unique, it is recognized that some degree of classification is possible. In similar environments the same forces tend to be operative to produce similar forms; geomorphologists can accordingly seek general explanations which may be widely applicable. It is fortunate for British geomorphologists that many of these environments and forces are present within Britain. Small though the country is, it is remarkable for the great variety of landforms which it presents to stimulate the geomorphologists' interest—'infinite riches in a little room'. Although it lacks Tertiary mountains such as the Alps and the Caucasus, and also landforms peculiar to the arid lands, it nevertheless concentrates a wealth of phenomena for study in its slopes, erosion surfaces, river profiles, river terraces, drainage patterns, and coastlines. The fact that the ground we tread is so varied, that it provides so much essential to economic enterprise, that it eases or impedes settlement, that it influences land use and affects the means of transport; all these emphasize how important and necessary it is to learn about the features of man's environment with which the geomorphologist is concerned. On the one hand, he helps us to understand regional and local changes of scenery by telling of the processes which have co-operated in the fashioning of mountains, plateaux, hills, lowlands, valleys and coasts. On the other, as an applied scientist, he can provide reliable data of practical value such as the physical changes taking place along our coasts, or gulleying and soil erosion for which human misuse of the land may be responsible.

Geomorphology tries to explain how the earth's surface has been sculptured by natural processes. This involves consideration of the forces at work, operating either from beneath or on the crust. Energy applied from within the earth affects vertical or horizontal movements of parts of the crust, creates mountain systems, shatters rock masses, and, no less dramatically, causes the emission of molten rock at the surface. While taking note of these internal, or endogenetic, forces, the geomorphologist is more particularly concerned with the exogenetic forces at work on the earth's surface itself. These are mainly the forces of erosion, themselves very much dependent on climate. The waves and currents of the sea, the water of rivers and lakes, snow and ice as they move down valleys or slopes, the winds, especially in the arid lands, and the force of gravity—these are the forms of energy which, by removing, transporting and depositing rock debris, are continually at work in fashioning the land. And since erosional processes have been at work throughout geological time under conditions that have been inconstant because of climatic and other physical changes, geomorphology is necessarily involved with chronological stages.

The methods employed by the geomorphologist are manifold and continuing to extend, marking the vigour and relative youth of this branch of geography. Direct observation has played, and must play, a great part. To the trained eye and imaginative mind it can clearly be rewarding. It was to the direct observation of James Hutton in the late eighteenth century that the science of geography, from which geomorphology can never wholly dissociate itself, owes its birth. Hutton noted—in defiance of current Biblical views of the earth—that the key to relief and landforms lay in operative physical processes for which he could conceive neither a beginning nor an end. However, direct observation has sharp limits set to it by the scope of human eyesight and the difficulties of extensive travel, although here modern aids, notably the helicopter, could be invaluable.

Direct observation is now supplemented by a wealth of material which can be analysed in the laboratory, notably samples of rock material, soils and pollen collected in the field which may contain clues to matters of wider interest. Air photographs, specially prepared morphological maps, geological and topographic maps and even historical maps, also repay scrutiny. A theoretical approach lies in the study of the physics of water, snow and ice and of their movements under varying conditions. Such an approach leads to the elaboration of models or of mathematical formulae that may help to explain and generalize particular cases. Accurate measurements can often be made—e.g., of slopes or of the rate of flow of a glacier—and statistical techniques employed, thus permitting precise statement.

All in all, geomorphology can now shed much light on the landforms of particular areas and how they have evolved; some at least of this is needed by regional geographers. Nevertheless only a small fraction of the earth's surface has been closely studied as yet, and geomorphology still has a long way to go. Much has still to be discovered, and the results offered frequently stimulate keen discussion among experts. Carefully assembled facts are gradually making geomorphology a more exact science, but workers have great need of imaginative power for interpretation; generalizations such as W. M. Davis's cycle of normal erosion mark stages in knowledge rather than final truth.

Political geography, like geomorphology, is concerned to study one aspect only of the world. The geomorphologist is concerned with phenomena which result from the operation of physical forces and have been only marginally affected by human agencies. In contrast, political geography focuses attention on essentially man-made structures, states, which pattern virtually the whole of the surface of the earth and are clearly significant elements in all environments.

A state consists of a land area occupied by a politically organized population. It is a complex human contrivance, sometimes relatively new, although often long established, and it is studied by workers in many academic fields. That the geographer is included among these springs from the fact that states are, *inter alia,* areal phenomena. To be more exact, states cannot exist without a territorial base, and the territory of a state is made up, not only of a land area, with a precise location, but also the air space above it and often a stretch of sea beyond its coasts called its territorial waters. A state is beyond doubt a most sharply defined region since its limits—international boundaries—are linear and, in most cases, delimited on maps and demarcated on the ground. The territory of a state presents familiar subject-matter susceptible to geographical analysis, and it is from this standpoint that a geographer's work can contribute to both the internal and external aspects of states. While on the one hand, political geography can have an applied value, by throwing light on internal political problems as also on international politics, on the other hand it offers data which cannot be ignored in holistic geographical studies.

A range of methods is employed in political geography, as in other systematic branches. Investigation and observation in the field are certainly relevant in the case of proposed changes of boundaries when it is necessary (in democratic countries at least) to discover the political attitudes of the populations involved. The laboratory offers little help, since the material for study are societies. Rather the well-stocked library, with its wealth of statistical, legal, historical and much other material is needed to provide relevant data for analysis. Numerical and statistical techniques can be used with effect, yet many problems face the student with imponderables, as when he tries to discover whether a particular state embodies a state-idea or merely a *raison d'être,* or when he tries to assess its social homogeneity, strength and viability. The method of analysis of the factors involved in a political situation or problem must be chosen carefully to make the approach as objective as possible.

The political geographer, like the geomorphologist, need never face unemployment, for problems abound and their disciplines are only at the active youthful stage. Political geography is developing its conceptual tools, faced with the difficulties of presenting in words, maps and diagrams, mental images of the world about us, a world subject to incessant change. And it need hardly be said that special difficulties arise for the social scientist, given the intractable complexity of his subject-matter. The quest is on, exciting and rewarding to disciplined work. And it is a geographical quest, from whatever vantage point one may set out, for it seeks the better understanding of the home of mankind.

FURTHER READING

BARRY, R. G. and CHORLEY, R. J., *Atmosphere, Weather and Climate* (Methuen, 1968).

CHISHOLM, M., *Geography and Economics* (Bell, 1966).

DICKINSON, ROBERT E., *The Makers of Modern Geography* (Routledge & Kegan Paul, 1969).

DURY, G. H., *The Face of the Earth* (Pelican, 1959).

EAST, W. GORDON, *The Geography Behind History* (Nelson, 1965).

EYRE, S. R., *Vegetation and Soils* (Camelot Press, 1968).

HARTSHORNE, RICHARD, *Perspective on the Nature of Geography* (Association of American Geographers, 1959).

8

Regional Geography

K. C. Edwards, C.B.E.

Regional geography has long been accepted as one of the principal forms of geographical study. To many in fact it is the means of attaining the geographer's primary objective, the understanding of the differences between one part of the earth's surface and another. The development of regional geography, though not a prominent aspect of the subject during its formative period in the nineteenth century, can nevertheless be traced to a much earlier date. Varenius, a German geographer in the seventeenth century, made a clear distinction between general (or world) geography and special (or regional) geography, while almost a century later Buache in France, seeking a convenient yet meaningful unit for study, smaller than the earth as a whole, proposed that geographical work should be based on river basins, an idea which has persisted in various forms to this day. Thus a study of the Nile basin in many of its aspects is essential to an understanding of social, economic and political conditions in Egypt, Sudan and Ethiopia as individual countries; in north-west U.S.A. the Columbia-Snake basin is often dealt with as a regional unit, while the Amazon and Congo basins invite similar treatment.

Regional geography is the complement of systematic geography, for instead of dealing separately with the different aspects of man's environment on a world scale, it is concerned to show how these elements interact together within individual areas on the earth's surface. After all, in the real world the various components of the environment (physical, biotic and social,) do not exist and operate separately. The growth of regional geography was in part a response to the fact that general or world geography, as pursued in the systematic branches of the subject, does not adequately explain the complex and often subtle differences which characterize individual parts of the world. It was also the consequence of a new attitude affecting the philosophy of geography itself. Until the closing decades of the nineteenth century the main purpose of the subject was to investigate the influence of the physical environment upon human communities, not that this necessarily meant viewing the matter from a crudely deterministic standpoint. It *did* mean, however, that the physical environment on the one hand and human or social conditions on the other were regarded as completely separate phenomena.

Pioneers in regional geography

Aided by advances in the social sciences, especially in sociology and economics, which affirmed the relative independence of man in relation to his physical environment, a new approach to the problem became evident. This approach was initiated by the French geographer, Paul Vidal de la Blache and also by Alfred Hettner in Germany. To Vidal, man makes adjustments to the physical conditions in which he finds himself and he in turn, by his actions, makes an impact on the environment, modifying it in many ways, even in the course of time being able to change it entirely in some aspect or other. Thus man and environment are really inseparable and the interaction of the two produces differential effects from one area to another. Moreover, a significant part of man's environment is man himself—man in society—or the social environment. This in essence is how Vidal perceived the man-land relationship. It provided him with a basis for the investigation of individual parts of the world as distinct from treating the earth as a whole. In his *Tableau de la Géographie de la France* (1903) he demonstrated the regional approach to the study of his own country by tracing the complex relations between community and environment which promote the geographical individuality of each significant part.

Vidal's work inspired many other geographers, most of all his disciples in France, to adopt the regional method. Besides Hettner in Germany, Mackinder and Herbertson in Britain also recognized its value, and before long it became generally accepted as a valid and effective means of treating the subject. In fact Mackinder, throughout his long career, repeatedly urged the importance of regional geography and first demonstrated its value in his book *Britain and the British Seas* (1902), which actually preceded the appearance of Vidal's *Tableau*.

Broadly contemporary with Vidal, A. J. Herbertson at Oxford was also concerned with the problem of dividing the world into smaller and more manageable units for the purpose of accurate description. In 1905 he introduced the idea of natural regions defined on a basis of climate, relief and vegetation, although climate was the primary determinant. Herbertson's contribution differed from that of Vidal in two main respects. First, it was based on the interrelationship of certain physical or natural elements, omitting reference to man; and second, it started with the world as a whole, which was then subdivided into its major natural regions. Vidal on the other hand was concerned with demonstrating the distinctive character of individual regions, assuming, by using both physical and human criteria, that no two areas are alike. He was not primarily concerned with developing a world system of regions.

Reference to these two pioneer attempts to establish a basis for regional geography, both of which exercised a powerful influence upon the development of the subject in their day, serves to indicate the main objective of the regional treatment. It is one way of organizing the wide range of facts with which the geographer must deal, the task being to discover how, on a rational basis, areal differences can be identified and the areas themselves described and delimited. Looked at in another way, the region is a device which the geographer employs to assist him in one of his most perplexing problems. Since every place and every small tract of land has its own distinctive features, it is impossible to describe and explain fully each small part of the world. Likewise, the historian finds it impossible to describe events through time, day by day or even year by year, so that he is faced with a similar problem and also uses a device to assist him, which he calls the *period*. This enables him to deal with the main events and characteristics of a longer span of time. The historical *period* is really a form of generalization. In like manner the geographer tries to treat together, within a framework of *regions,* the significant characteristics of many places. The *region,* too, is a form of generalization and is therefore strictly an intellectual concept. As K. B. Cumberland has stated. 'a region is not an object that exists, but a device for facilitating the description and understanding of the differences that are found from part to part over the earth's surface.' This is not to say that all regions are non-existent, as we shall see later.

There are many different ways of applying the use of the region. These vary with the particular purpose for which the regional treatment is intended. The purpose might be to subdivide the world into physical units of some kind, such as Herbertson's scheme of natural regions; or it might be, to take a more modern example, to examine the connections between a large city and the surrounding country with its smaller towns and rural areas, in order to define what is called the city region. Obviously the criteria used to identify regions will vary according to the kind of unit sought. However, since the region is fundamentally a form of generalization, it follows that any portion of the earth's surface so designated should possess a high degree of uniformity with regard to those qualities or characteristics which serve to distinguish it from other areas.

Both in teaching and research the scale of regional subdivision depends on its purpose and on the degree of detail required. Thus for the intensive study of very small areas (termed micro-regions), for very large areas dealing with general global matters (macro-regions), and for those of intermediate size (meso-regions), the scale would vary accordingly. This indicates a hierarchy of regional scales, the choice being determined by the purpose and character of the study, which is a very different matter from a hierarchy of regions. In some circumstances, however, as in a survey of landscape types, a system of regional divisions devised on the hierarchic principle can be achieved.

Generic or formal regions

It is customary to recognize two main categories of regions —generic and specific regions respectively. Generic, or formal, regions are derived from the systematic branches of geography and represent the spatial expression of particular aspects of the environment: physical, biotic and social. Examples are climatic regions, morphological regions, vegetation regions, land-use regions, linguistic regions and a host of others. These are generally single-aspect or single-element regions, though they may be based on two or even more elements if a distinct causal connection exists between them, for in such cases there is likely to be a broad conformity in their distributional patterns. Since formal regions, like their respective branches of systematic geography, relate to the world as a whole, they represent types which occur within each element treated. Thus, although each region is uniform in respect of a given element, the world distribution pattern reveals a repetition of types, as for instance the recurrence of areas of the Mediterranean type of climate on a map showing climatic regions.

In general, world patterns of formal regions differ one from another according to the aspect or element treated. Such regions, or systems of regions, are of value not only in revealing the repetition of types but also in comparative studies dealing analytically with the various components of the total environment. For it must be recognized that formal regions, while they can be described and defined, do not exist in isolation. Here it is of interest to mention one of the most familiar of all the maps commonly found in an atlas, which is described as 'The World: Political'. It depicts in different colours the independent nation states, the self-governing units of the Commonwealth, colonies, dependencies and so forth. Dealing with the single aspect of political status, this is clearly a map of formal regions in which we again note the repetition of types, in this case degrees of political independence; and if the key were sufficiently detailed we could note the repetition of types of government or constitution: monarchy, republic, federation. At the same time, with regard to individual countries, especially the nation states with their differing languages, cultural traditions and economic conditions, this map delimits non-formal or specific units, in each of which the interaction between many different elements is so complex as to render each unique, save for the one common link of national independence.

The elusive specific region

While the derivation of the generic or formal region is beyond question and its nature readily understood, it is quite otherwise with the specific region. For this concept is the subject of prolonged controversy which from time to time flares up as a topic of sharp debate among geographers, making it a contentious matter on which agreement remains elusive. Since the purpose of regional geography is the understanding of the earth's surface through the study of its constituent parts, geographers have inevitably sought to find a basis for regional work which will include both the physical and human components which operate together, interactively, to form the total environment. Formal regions are inadequate for the purpose since they deal separately with the individual aspects which cannot be treated in combination in respect of any one area. Natural regions, which may be regarded as formal regions combining one or more physical elements, are equally unsatisfactory since they exclude reference to the all-important social elements. It is because individual parts of the world differ from one another in respect of their total conditions that the concept of the specific region arises. Such areas are specific in the sense that each is unique, for, as many writers have pointed out, geography, like history, never repeats itself in detail. The problem is therefore a dual one—of discovering an acceptable basis for this form of region and of devising the appropriate technique for delimiting it.

Many geographers have attempted to find solutions to this problem but none has succeeded in doing so on strictly objective lines. Always, to a greater or lesser extent, subjective judgement has been involved in deciding upon and assessing the criteria selected to provide the basis, and because of this most of the attempts take the form of description within a predetermined framework. In as much as geography is not an exact science, but is essentially interpretive, complete objectivity may well be impossible. This is not to say that the results are necessarily inaccurate and, in any case, accuracy and objectivity are not synonymous. However, the dilemma presented here is sufficiently acute for some geographers to reject the case for the specific region altogether and to abandon the quest for it. E. W. Gilbert and others who follow the Oxford tradition acknowledge the difficulty but regard geography not as science but 'as the art of recognizing, describing and interpreting the personalities of regions.' (see 'The Idea of the Region', *Geography,* Vol. 45, July 1960).

In view of the difficulties it is instructive to examine briefly some of the ways in which geographers have tried to find a satisfactory answer to this question. For no one would deny the existence of certain specific regions. Those parts of our own country which are known as East Anglia, the North-East, South Wales and the Central lowlands of Scotland are recognized as unmistakeably distinct regions. Each of them exhibits a particular assemblage of features not shared by the others. The features themselves, physical and human, are readily observable, their interaction can be demonstrated, and at least within broad limits a boundary can be drawn. The commonest approach to the specific region is through what are termed its distinctive characters. These refer to features of the landscape, both physical and cultural, which combine to make it different from others, so that the boundary of the region is the limit of the occurrence of such features or a combination thereof. The method involves landscape analysis, the identification of characteristic features, an evaluation of the interrelations between them, and a concluding synthesis which portrays the overall nature of the region. The latter signifies the individuality of the region, or its 'personality', as Vidal went so far as to call it and as Gilbert has recently reiterated. While this method has the advantage of providing the basis for an ordered description, it frankly attains to little more than a description. Nevertheless, many textbooks have been modelled on this approach, one of the most satisfactory being *New Zealand, A Regional View,* (1958) by K. B. Cumberland and J. W. Fox, in which flexibility of treatment resulted in due emphasis being given to the dynamic effects of cultural factors in distinguishing the regional divisions.

Some geographers have concerned themselves with what is called the geographic region. This term is used to imply something more fully geographical than the generic region and less artificial than the political unit. The keynote to this type of region, as stated by J. F. Unstead, one of its earlier advocates, is the integration of natural and human factors, a notion which clearly differs from the idea of the region identified by its distinctive characters. Unfortunately this holistic approach is impracticable for, as we have already seen, the patterns of individual physical elements seldom conform with one another, while physical and human patterns conform even less. An equally important objection, which incidentally affects regional geography in general, is the contrast between the rates of change in physical and human conditions respectively. Changes in the latter are rapid, and the cultural impress on the landscape is in some parts continually shifting, while in others it is becoming more and more intensified. One of the most cogent criticisms of the theory of regions is that many forms of the concept result in units which are too static, so that sooner or later they lose their significance and even their validity. This is why the regional concept itself should take full account of the dynamic of human affairs, and it is also the reason why the pioneer work of Vidal at the beginning of the century has long been outmoded, partly because of later advances in theory and techniques, but even more because of changes in the social and economic context, which make the approach irrelevant and inapplicable to present-day conditions.

Another contribution was made some years ago by the American geographer Derwent Whittlesey. Rejecting the idea of the region based on a totality of elements, he favoured a selection restricted to those which are functionally associated with man's occupance of the earth. The geographer is left to select and to emphasize the elements according to the nature of the area treated. As a unit the latter is termed not a region but a compage, an obsolete word meaning structure or composition. The compage is a somewhat vague and subjective idea but has the merit of flexibility, especially as the emphasis based on one or more of the elements allows for the expression of a central theme in which the real interest of the particular area may lie. In this respect the notion of the compage, despite its subjectivity, was at least an attempt to liberalize the concept of the region, to free it from a conventional mould and to invest it with something more than description. In his large-scale study *Latin America* (1941) Preston E. James, another American geographer, went even further. His approach to regional divisions was on a basis of leading themes, mainly relating to population and the economic and social conditions of human occupance. By adopting an historical-cultural viewpoint, James discussed fully only those aspects of the physical environment, usually relief and vegetation (with climate), which appeared relevant to his purpose. The result was a vivid and stimulating account of the major divisions of the continent within the framework of the different countries.

The functional region

A fruitful approach to the idea of the specific region is to place the emphasis fairly and squarely on human activity since man is progressively achieving mastery over his physical environment. This is especially true of those parts of the world which exhibit well-organized agriculture, modern industrial complexes and large cities. Such areas are characterized by a unity based on an economic and social organization which reflects the interconnections of places within them. A region identified thus in terms of organization is called a functional region. The Swiss geographer Hans Carol has expressed the contrast between generic regions and regions of this type, by referring to them as formal and functional regions respectively. An example of the functional region is the area served by a large city defined as the territory over which its distributional functions are spread. While the spatial limits of these functions will not coincide, many of them will do so sufficiently well to delimit the region with reasonable accuracy. An extension of this idea to the case of the very large city having the characteristics of a metropolitan centre and serving as the integrating focus for a correspondingly large area provides the concept of the city region, which many people favour as a realistic unit for a reformed system of local and regional government.

A good illustration of a functional region in Northern England is cited by R. Minshull who points out in *Regional Geography: Theory and Practice* (1967) that while the industrial lowlands of Lancashire and the adjacent hills of the western Pennines belong to two distinct formal regions as regards configuration, functionally they both form part of a single region. The upland, supplying the lowland with limestone for the chemical industry, water for cotton mills and urban populations, livestock for farms, and open country for recreation, forms part of the one organizational entity. Its closely-knit inter-connections provide its distinctiveness rather than any special features of landscape. The region moreover, unlike one of the formal type, is neither uniform nor homogeneous but instead derives its unity from intrinisic diversity. Again, in the U.S.A., the area administered by the Tennessee Valley Authority, involving parts of several states, may also be regarded as a functional region, for despite the variety of terrain and of the man-made features promoted by the great multi-purpose project, it is the integration of the various functions through organization, which gives it unity. These examples indicate that the functional region is a dynamic concept. It is based on the operational connections between phenomena and places and, unlike other kinds of region, is not primarily a descriptive generalization and to that extent is a more sophisticated concept.

The limitations of descriptive regional geography

From what has been said in this brief review of the nature

of regions, it will be evident that most of the attempts to formulate a basis for regional treatment have relied upon finding a satisfactory method for description. In itself this is not inconsistent with the general aim of regional geography which is to promote an understanding of the individual parts of the earth's surface. In teaching at an elementary level, descriptive regional geography based on simple correlations between associated phenomena has its place. At a more advanced stage, however, it fails to satisfy intellectual curiosity, for description alone neither advances understanding nor promotes new knowledge. For research it is entirely inadequate. Added to this is the fact that purely descriptive work becomes monotonously repetitive and eventually sterile. The failure to recognize this situation in the past is one of the reasons why in some quarters regional geography has fallen into disrepute. This is unfortunate because, with the current changes taking place in geography as a whole, the scope for regional work, instead of diminishing, has greatly increased. Moreover, it is in its dynamic rather than its descriptive role that the advance is already taking effect.

New directions in regional geography

The reinvigoration of regional geography can be attributed to three separate influences. The first of these is the effect upon the content of geography of the profound and widespread changes in the modern world consequent upon the Second World War and its aftermath. The second is the expanding contribution made by geographers to the study of practical problems in the fields of both the physical and social sciences, a development generally comprehended by the term 'applied geography'. The third influence at work, one of a different kind, is the increasing use made of statistical data and of quantitative techniques in dealing with them, which result in greater objectivity and precision in regional research.

The far-reaching changes affecting the contemporary world include the adoption of communism in its varying forms in a number of countries, paralleled to some extent by the growth of state intervention in economic and social affairs in capitalist countries; the world-wide process of de-colonialization and the emergence of newly independent states; the new groupings of countries for mutual economic advantage; and the revolution in methods of transport and communication. All of these have given rise to new geographical facts and to new attitudes affecting their spatial relationships. In the previous chapter, attention was drawn to the benefits resulting from the penetration of systematic geography into the scientific borderlands. So, too, the surge of cultural change in the modern world affords a fresh incentive to regional geography. Not only are the functional region and possible variants of the thematic region suitable as devices for the study of present-day problems in their spatial context, but many new spatial groupings have come into existence which demand a regional treatment. Some of these occur on a macro-regional scale and present increased opportunities for geographical study. The territories forming what is now termed South East Asia, for example, must be recognized as an important culturally coherent unit within the humid tropics. Similarly the lands bordering the south-west Pacific, including the island clusters, which have undergone immense changes in political status, strategic orientation and economic development, form another macro-region. The south-west Pacific is a reality, already grasped by people in Australia and New Zealand, though less readily by us, which claims the geographer's attention as a huge oceanic domain, fast becoming a functional entity through modern communications and the application of economic and technical aid.

Closer at hand in Europe, the six countries of the Common Market present a grouping of states which, as an economic entity, gives them collectively greater prominence as a regional unit than as separate countries. Certainly it makes greater sense today to treat Italy, a modern industrialized member-state, as a component of the E.E.C. rather than as a country of a nostalgic Mediterranean world. The five mainland states of the Caribbean which constitute the Central American Common Market (C.A.C.M.), also having economic integration as its aim as distinct from a mere free-trade association, provides yet another significant unit for study. Many other parts of the world having a new significance under present-day conditions afford scope for investigation and re-appraisal by the regional geographer.

Turning now to the geographer's part in the study of current problems, we distinguish another direction in which his capacity to deal with variable elements of man's environment in their area context allows him to make a contribution. The demands of society in a rapidly changing world relate in many instances to a re-shaping of the cultural and to a lesser extent the physical habitat. The problems of urban renewal and expansion, industrial location, amenity provision and nature conservation in the older advanced countries, equally with problems of food production, resource development and population control in the newer developing countries, can only be satisfactorily solved by reference to a policy or plan. Their solution in turn involves the use of land, often a change of use, and even if the problems themselves occur on a nation-wide scale, the measures adopted to deal with them affect particular areas. Thus ultimately they become locational problems, and the geographer's ability to synthesize is clearly relevant to their study.

One of the most important applications of the regional principle to present-day problems is to be found in the growing movement for regional planning. This is the concern of economists, sociologists and other specialists besides geographers, since it involves the wide field of spatial organization. It reflects nevertheless a tendency towards convergence between geography and the other social sciences, for the problems requiring investigation are often best approached from an interdisciplinary standpoint. Naturally enough regional planning in communist countries, such as the U.S.S.R. and Poland, emphasizes economic planning through 'economic regionalization', to use the cumbersome phrase describing the process; but in Western countries too, notably in Britain and France, national economic policy is now being applied on a regional basis. In Britain the regional framework for this purpose was set up by the Department of Economic Affairs in 1964 and, except for some modifications, closely resembles that of the earlier

Standard Regions. Each unit consists of a group of counties with a large city serving as the regional centre. This adherence to administrative divisions is frequently contested by geographers on the grounds that they are geographically unreal. Yet there is an important principle involved in their use for, as recently noted by D. Turnock, in 'The Region in Modern Geography' (*Geography,* Vol. 52, November 1967), the relationships between population and resources of each country are primarily organized at state level though variations occur from one part to another, giving rise to individual regional problems such as unemployment, defective communications and derelict land, which may of course transgress regional boundaries. Interregional arrangements, however, can usually be made to deal with such cases. Furthermore, the availability of many kinds of statistical data which normally relate to administrative units, is an advantage almost sufficient in itself to justify this basis for regional delimination. The fact, however, that in many cases these divisions no longer reflect the existing facts of population groupings, economic activity and many other items of social organization, is a serious disadvantage and is the essential reason for the geographer's objection to their use. The fact that administrative divisions are not directly related to physical conditions is on the whole of little consequence.

The division of a single country into major units for the purpose of implementing policy, whether official or unofficial, is called regionalism. Regionalism takes many forms, varying from the territorial units which are found convenient for the administration of organizations such as the Automobile Association (private), the B.B.C. (semi-official) and the National Coal Board (official), to the idea of developing some of the powers of government from the central authority to provincial units. In Britain the demand for autonomy by the Scottish and Welsh Nationalist movements represents an extreme form of political regionalism. On the other hand, the need for devising new administrative areas, larger than the existing counties, is now generally accepted. The case was first examined by C. B. Fawcett as long ago as 1917 and subsequently published as *The*

Provinces of England (1919, new edition 1960). This work, though not characteristically an essay in regional geography, was assuredly a pioneer study in applied geography.

The third influence affecting modern regional geography, the increasing use of quantitative techniques, is mainly effective in studies on a micro- or meso-regional scale. The application of acknowledged mathematical procedures to statistical data not only secures greater objectivity and precision in geographical work but widens the scope for analytical studies dealing with multiple factors or variables in spatial expression. For example the treatment of a complex of geographical variables by different methods of multivariate analysis such as factor analysis, helps to determine the character and extent of the core of a region.

With regard to the investigation of phenomena within a region, two distinct forms of analysis are available, the one being termed static and the other dynamic. The former is essentially descriptive of the distribution of variables within a region. Among the static methods correlation techniques indicate how closely (or not) the values of one variable are related to those of a second or third or even more variables. Simple regression techniques may show how change in one variable is affected by change in another. Trend surface analysis, an extension of the regression technique, makes possible the measurement of the relationship of one variable to the two map grid co-ordinates. In its simplest form a plane is fitted through points located in respect of three values. The techniques for dynamic analysis measure processes occurring with a region. For this it is necessary to assume certain geographical relationships and to test the assumption for various times, simulating real conditions. An example of this approach is the gravity model which assumes that movement, e.g., traffic flow, between two towns is directly related to their populations multiplied together, divided by the square of the distance between them. This formula can be tested and used to describe and even forecast such a process. Clearly such techniques have opened up new opportunities for work in depth, and it is largely through such tools that regional geography is being adapted to modern needs (see P. Haggett, *Locational Analysis in Human Geography,* 1965).

Although from the methodological standpoint there are obvious grounds for distinguishing between systematic and regional geography, in practice the separation of the two tends to become less marked. In fact, the more dynamic regional geography becomes and the more it concerns itself with the study of problems within an area setting, the more this distinction becomes unrealistic. At an advanced level of study, and in research, the investigation of problems is more important than preserving the sanctity of the region in the traditional sense. Consequently the regional method, which involves the integral treatment of components in a spatial setting, is more fundamental to geography than merely seeking the delimitation of a regional unit. For geography is essentially an integrating discipline, and some form or other of the regional method can best ensure the fulfilment of that role.

FURTHER READING

COLE, J. P. and KING, C. A. M., *Quatitative Geography* (John Wiley, 1968) pp. 631-9.

DICKINSON, ROBERT E., *The Makers of Modern Geography* (Routledge & Kegan Paul, 1969).

HARTSHORNE, RICHARD, *Perspective on the Nature of Geography* (Association of American Geographers, 1959).

JAMES, PRESTON E., *Toward a Further Understanding of the Regional Concept* (Annals of the Association of American Geographers, Vol. 42, September 1952).

McDONALD, J. R., *The Region: Its Conception, Design and Limitations* (Annals of the Association of American Geographers, Vol. 56, September 1966).

ROBINSON, G. W. S., *The Geographical Region: Form and Function* (Scottish Geographical Magazine, Vol. 69, 1953).

WRIGLEY, E. A., 'Changes in the Philosophy of Geography', in *Frontiers in Geographical Teaching* (eds. Chorley and Haggett), (Methuen, 1966).

9

Land Use and Resource Analysis

Alice Coleman

The purpose of this chapter is to outline one of the central themes of geography which illustrates how the various aspects discussed earlier, physical and human geography, systematic and regional geography, are drawn together and interwoven. The theme is man's adaptation to, and use of, the land and other resources.

Fundamentally the earth possesses only two resources: space and time. Everything else that may be thought of as a resource can be resolved as a component of one or both of these. Space is a continuum of land and water, atmosphere and biosphere, which presents a rich and varied range of resource opportunities, and conversely, resource limitations. The contrasts which it exhibits, even within a few hundred yards, may be so great as to make all the difference between prosperity and bankruptcy, if both sides of the line are used for the same purpose. Consider the case of two early settlers in the St Lawrence Valley clearing forest on adjacent land grants to establish farms. One, revealing level fertile postglacial clays, found a soil resource capable of responding to all the demands he made of it, while the other found sterile and rocky soils of the Laurentian shield incapable of responding to similar demands. The poor quality soil kept the second settler in a permanent state of poverty and finally drove him away to seek better fortune elsewhere.

The time resource, too, is available in varying quality.

Once again consider two farmers, both of whom spend the same number of man-hours tending their fields. One is an Indian peasant, whose time input results in sufficient food to support his family. The other is a New Zealander, who produces enough food to feed thirty families. It is not the quantity of time available that is different but its quality. The New Zealander's time is a better quality resource largely because he has the knowledge and the capital to reinforce his own human energy with other forms of energy such as machinery and fertilizers.

Geographers are interested in these quality variations of the space resource and the time resource. They want to explain them as individual cases, and also to organize the individual explanations into patterns and frameworks of general applicability. In exploring the reasons behind the phenomena geographers often find that they need insights into disciplines other than their own. They must therefore be outward looking and intellectually adventurous in exploring the borderlands of their subject.

The broad general explanation that geographers seek seemed to be within reach in the nineteenth century with the concept of geographical determinism. This is the idea illustrated by the St Lawrence Valley example cited above, namely that there is a correlation between the favourable or adverse nature of the environment and the socio-economic condition of the inhabitants. This is tantamount to saying that poor quality in the space resource at any place determines that there should also be poor quality in the time resource at that place, i.e., that the environment is man's master and not vice versa.

It is easy to understand the intellectual appeal of geographical determinism. In the nineteenth century new lands were being explored and a bewildering wealth of new material was being added to the subject. Determinism appeared to explain why many of the newly discovered peoples were existing at such primitive economic levels. The barren Australian desert, the enervating African rainforest, and the frozen Arctic tundra were all adverse environments that harboured primitive peoples, and determinism was reinforced by the fact that Europeans frequently deteriorated if they lived for too long in such regions of difficulty.

At the time of their discovery many primitive tribes had been living for centuries in a state of cyclic balance with their environment and the deterministic view was often essentially valid. However, by the time the theory was being formulated and elaborated, its reality was already beginning to wane. Rapid changes were sweeping the earth as more advanced economic groups began to occupy the same areas as primitive tribes and yet maintained themselves at much higher economic levels than the original inhabitants. Furthermore, the tribes themselves began to be insulated from the adversities of the environment and especially from the diseases endemic in it, so that they could no longer live in the traditional way, even if they had so wished, because their numbers had increased beyond the scope of the previous economic balance. Instead they availed themselves of the new knowledge to change the quality of their time resource, and through it the quality of their space resource. The deterministic model which had served fairly well at first now became a frame of reference for detecting departures from the model, and these increased in number until the model itself came to be discredited.

The weakness of determinism in its most extreme form lay in its assumption that specific correlations would never change. If man was limited by his environment at a given time, the same environment would exert the same limitations indefinitely. But this is not the case. The space resource is not the only independent variable; the time resource can also vary independently and produce an entirely different equation.

This can be exemplified by the area formerly known as the White Highlands in Kenya. In the late nineteenth century, Kenyan occupancy was typically deterministic. The most favourable areas were those that received enough rainfall to support peasant cultivators. The next most favourable received less rainfall but enough to support nomadic herdsmen. The least favourable were the driest or coldest areas and these were uninhabited. But although the Africans regarded the uninhabited lands as the poorest quality space resource the British saw them as an opportunity. They settled in large estates, supplemented the rainfall with ground water and upgraded the land to a highly productive condition. Since national independence parts of the White Highlands have been taken over by native Kenyans who bring to them a better quality of time resource than their forbears did, and consequently see these lands as a better-quality space resource.

As man improves his use of time, his power over the environment increases together with his choice of land-use alternatives. For a while this idea was fashionable in its extreme form, namely the principle of possibilism which regards the opportunities for environmental control as virtually unlimited, i.e., that man is the master of the environment and not vice versa. Today's geographers have retreated from this position also. They recognize three very definite kinds of limitation in environmental control. The first is physical or technical, in which it is not yet known how to produce the desired control. The second is economic, when suitable techniques exist but are too expensive to implement. The third is cultural, when technically and economically feasible projects are eschewed on account of cultural attitudes such as lack of education, religious beliefs, or conflict with other values that are prized more highly. Religious attitudes to cattle in India and the rejection of Stanstead as a site for London's third airport are somewhat dissimilar examples of this third category.

Probabilism, the intermediate position which modern geographers adopted, recognizes that both the space resource and the time resource may vary in relative dominance, and that they interact together to produce an integrated complex, the use of land, which is one of the most prominent aspects of spatial variation on the earth's surface. Probabilism inherits from possibilism the positive attitude that the space resource can be improved, and from determinism the realization that limitations exist and can be intensified by unwise land use. Hence it combines ideas of development and conservation alike. It is also tailor-made for the application of probability mathematics, which are gradually making land-use studies more precise and orderly.

Systems of land use fall into two basic types: ecosystems and economic systems. The former were dominant throughout the greater part of human development but are much more subsidiary today. They are studied as being of intellectual importance in providing a number of significant reference points, past, present and future, that help us to understand economic systems of land use better.

A simple but classic illustration of an ecosystem is derived from the records of the Hudson's Bay Company, which noted a marked periodic fluctuation in the number of lynx pelts brought in by trappers. Every decade or so in the nineteenth century there were cyclic fluctuations with minima around 3,000 and with maxima around 63,000. The basis of this cycle was a simple food chain, since rabbits formed the food of the lynxes. Starting at the point where the rabbits began to multiply, the lynxes had access to a more abundant food supply and after a short time lag they, too, began to multiply. Both species increased until they were so numerous that the rabbits could hardly venture out without encountering a lynx and the lynxes had to make very little effort to obtain a meal. Rabbit numbers thereupon began to decline, and inevitably in time the lynx numbers followed suit until both species became few in number. At this point the rabbits became relatively free of their predators and could multiply again, thus initiating a new cycle.

Although a more typical ecosystem might contain hundreds of interacting species, this simple case is sufficient to illustrate four basic principles. The first is the existence of long-term stability, the balance of nature or homeostasis, while the second is a set of short-term fluctuations whereby any disturbance of the long-term balance is countered by a natural system of checks. Third, this self-correcting mechanism is made possible by high birth rates and high death rates which allow recovery from the critical low points and high points of the short-term cycles, respectively. Finally there may be occasional new elements introduced by mutation or migration for which the natural checks do not cater. These constitute transformations of the ecosystem, and transmit adjustments throughout its linkages until the new element has found a stable niche, possibly by extinguishing one or more of the pre-existing elements. In this way ultra-long-term evolution can be superimposed upon long-term stability.

When man first evolved about half a million years ago he was simply one more species in the primaeval ecosystem and conformed to the four principles. His long-term evolutionary role was to find an ecological niche in all the areas where he could survive, and there he existed in long-term stability with food supplies and predators, but also subject to short-term fluctuations generated by the various processes in the ecosystem, including his own activities. His condition must have been one of high birth rates and high death rates; our knowledge of stone-age skeletons suggests that the average expectancy of life was little more than twenty years.

The space resource at this time was completely unimproved wildscape: jungle, forest, prairie, desert, tundra, icefield, etc. Since man had to rely entirely on his own unaided efforts he was unable to inhabit many of these environments; the space resource was too poor for the time resource at his disposal. This can be related to the energy requirements of the human body which is itself a system, and like most systems depends on inputs of energy to balance outputs of energy. Failing this the action of entropy will bring about degeneration and decay.

The output of energy by early man had to be prodigious by modern standards, as he had to range over a very wide area to hunt and gather enough wild food to keep him alive. His input of energy was not only food; he also needed the warmth of the sun all the year round, which is why his original base can be traced to tropical Africa. Here there was not only sufficient warmth to act as a direct source of energy but also sufficient heat and moisture to support a rich and varied biosphere which furnished plant and animal foods all the year round. Furthermore, a sufficient supply was concentrated within relatively small areas which meant that energy input could more easily match energy output, as energy was not being dissipated in long forays.

About 10,000 years ago there began the major transformation known as the Neolithic Agricultural Revolution, which ultimately replaced the ecosystem with an economic system through the conversion of wildscape to farmscape. This was achieved by altering the energy inputs and outputs that had maintained the wildscape ecosystem. Settled agriculture meant a larger and more certain food supply, because both plants and animals had been brought under man's control. He had in fact improved the quality of his time resource by harnessing the energy of animal fertilizer and also by reducing weeds which competed with food plants. This increased his harvest, nourishing him better and supporting larger numbers.

The conversion of wildscape to farmscape is an example of possibilism, since man now had the choice of two systems. But it is also an example of determinism, since the locations where this choice could be exercised were restricted. Physical factors, which had previously differentiated wildscape habitats into habitable and uninhabitable types only, now began to create more numerous differentiations, since not all the inhabited wildscape was conducive to agriculture and some of the uninhabited was. For example, the tropical areas that favoured early man contained both the major enemies of agriculture-heavy leaching that rapidly destroyed the fertility of deforested soils, and rampant weed growth that competed too severely with cultivated plants. In such conditions agriculture could only survive by periodically shifting to new territory in a system of land rotation. In other areas dry conditions precluded crops, and nomadic herding was appropriate. Neither of these roving economic systems allowed the same accumulation of possessions and development of civilization as did settled agriculture, which is believed to have originated in two areas, the Middle East and Mexico, on the basis of wheat and maize respectively. That these areas should have come to be considered the richest space resource is an example of the change in geographical values that accompanies a transformation of the time resource.

A whole series of smaller transformations at long intervals of time allowed the twin problems of weeds and infertility to be reduced in influence so that a bigger harvest could be obtained from less land, with a smaller proportion left fallow. This permitted the progressive extension of farmscape into previously unsuitable wildscape and a progressive diversification of farming techniques to overcome the various kinds of physical limitations. The wildscape also became more diversified. As well as being uninhabited, or occupied by hunting and gathering societies, it was also used for extensive grazing or for lumbering, which modified the ecosystem but not to the extent of replacing it with created farmscape.

The creation of farmscape from wildscape is the replacement of an ecosystem with an economic system by means of a transformation. At first, and again with each successful innovation, there was an increased throughput of energy which allowed the land to support a larger population. The element of a high death rate was removed from the scene during the transformation period. But population increase could not be maintained indefinitely. After a time the innovation reached its ceiling capacity and the transformation entered into a limiting union with the ecosystem to form an economic system. The economic system could support a higher density of population than the primaeval ecosystem alone, but the ecosystem imposed on it cyclic fluctuations with high death rates at the new ceiling level.

This can be illustrated by conditions in mediaeval England. The late twelfth and the thirteenth centuries saw a population expansion accompanying economic growth based on widespread conversion of wildscape to farmscape. Once this transformation reached its ceiling, continued population growth meant poorer nourishment for each individual and lowered resistance to disease. Hence the ecosystem's cyclic checks came into play in the form of epidemics, which were much more severe than during the transformation phase. The most notorious was the Black Death of 1348 which so reduced the population that it did not regain its former level until the early seventeenth century when improvements associated with the enclosure movement had created a higher effective ceiling.

The eighteenth century witnessed the beginning of another major transformation in human history, namely the Industrial Revolution, which was comparable in importance with the Neolithic Agricultural Revolution. It also produced a characteristic habitat normally referred to as townscape. Clearly towns and industry had been known from a much earlier date, but before the industrial revolution they were subsidiary to, and dependent upon, the farmscape. After the industrial revolution the expansion of the towns began to dominate the landscape, eventually leading to the modern conurbation and megalopolis. Britain, the earliest industrial country, was the only one to have become predominantly urbanized by A.D. 1900, i.e., to have more than half its population living in towns of over 100,000 people. Today over one third of the world's population is urbanized by this definition. There has also been a tremendous growth in the number of cities with over one million people, from only one (London) in 1850, to five in 1900, eighty in 1950 and approaching 150 today. Furthermore the urban sphere of influence extends far beyond the townscape area *per se*.

The industrial revolution, like the neolithic agricultural revolution, was based on a great advance in the use of energy. Man now secured much greater command over inanimate energy sources, the forms of which have been steadily diversified during the industrial age. The variety of manufactured products has also increased, enabling progressively more people to live at a high standard of material comfort; and this, in turn, has stimulated a market for information, so that it has become more common to make a living out of industrial expertise, or out of education and research. None of this would have been possible, however, if the early industrial discoveries had not been accompanied by the development of the Norfolk four-course rotation, a most ingenious system of farming which introduced several mechanisms for combating agriculture's two perennial enemies, weeds and soil sterility. So successful was it that for the first time there could be continuous cropping, dispensing with fallowing practices. More food could be produced from less land by fewer people, thus initiating the release of farm workers to industrial occupations. This trend, subsequently reinforced by farm mechanization in the twentieth century, has been maintained ever since so that today only about 3 per cent of Britain's labour force is employed in agriculture while farm techniques and productivity continue to be improved. In fact, agriculture, too, has been industrialized.

In Britain the industrial transformation has not yet reached a ceiling where high death rates and cyclic fluctuations of human numbers would occur—a situation due in part to our heavy reliance upon the world economic system of international trade. Ceiling conditions, however, still obtain in some of the developing countries, but there they are more the outcome of the pre-industrial economic system that is being rejected than of the industrial economic system that is being introduced. In these cases the transformation is proceeding unevenly with some features, such as medical care, advancing more rapidly than others such as the efficient improvement of the food base. The resultant problems could be solved by full industrialization, including the industrialization of agriculture. The technical knowledge for completing the transformation is in existence; what is actually impeding progress is either economic limitations, namely a lack of capital to implement the technical possibilities, or cultural limitations, namely a lack of the mental attitudes needed to overcome the economic limitations. Economic and educational aid are helping to resolve these problems. Meanwhile the developed nations are learning how to offset future cyclic fluctuations in population by controlling birth rates instead of relying on high death rates.

A marked feature of the industrial age has been a vast increase in the variety of uses which man can find for his land and other resources. In the developed countries many hundreds of different land-use categories can be recognized for mapping. These are so intermingled in space that their analysis requires a higher order of conceptual organization, notably systems which are able to combine the various classes of use into comprehensive functional patterns.

A first approach to systems analysis of land use recognizes the three basic systems of wildscape, farmscape and townscape, which are capable of more detailed resolution into subsystems, and which can also generate transformations where they interact. Each of these systems may contain any or all conceivable categories of use. They cannot be defined on a basis of category inclusion or exclusion, but rather in terms of the dominance of functionally related categories. Wildscape is dominated by cover types, especially natural and semi-natural vegetation, but also rock outcrops, sand-flats, glaciers and water surfaces. Farmscape is dominated by ground crops and tree crops, hayfields and man-made pastures. Townscape is dominated by buildings, transport facilities and tended open space. Each scape territory can absorb subordinate quantities of uncharacteristic uses. For example, townscape absorbs patches of woodland or a few nursery gardens, farmscape absorbs farm buildings and small villages, and wildscape absorbs occasional reclaimed fields or isolated cottages. In small quantities these alien categories do not disturb the fundamental stability of the system.

The three basic systems are subject to essentially different controlling agencies. Wildscape is primarily maintained by nature, farmscape by the private planning of the individual farmer and townscape by the control of public authorities. Difficulties arise when a given controlling agency intrudes into an inappropriate system. Farmscape, for example, will suffer both when nature becomes too strong to contend with, and also from uncontrolled incursions of the general public from the townscape. Such difficulties may become chronic and widespread in fringe situations where the territories of two different land-use systems adjoin each other along a frontier zone.

There are two main types of fringe territory. Where wildscape adjoins farmscape there is frequently an intermediate zone of marginal land, and where townscape adjoins either of the rural scapes there may be a rural-urban, or rurban, zone. Neither of these can be defined by dominance of related uses. Instead they exhibit mixtures of unrelated uses in co-dominant proportions. The marginal fringe contains farm fields co-dominant with rough vegetation and/or woodland, and the rurban fringe contains urban sprawl co-dominant with fragmented farmland, often interspersed with idle land also. The fringes are often situations of conflict in which neither of the unrelated use-groups can function with maximum efficiency; both the component systems become unstable and degenerate into negative transformations accompanied by human hardship.

Thus the inhabitants of sprawling rurban settlement incur added expense in financing water and sewage systems that must include dead sections to bypass the fragmented farmland, while the farmers also incur added expense on account of damage by trespassers from the urban areas. Good planning attempts to eliminate the rurban fringe and to replace it by a 'fence' or line which separates farmscape and townscape more sharply and allows each to function without interference. A great deal of Britain's former rurban fringe has now been successfully replanned as compact townscape, and attention is also being given to the marginal fringe between farmscape and wildscape.

The fringes are mobile; they migrate across country as their bordering scapes expand or contract. If the migration results from expansion of the more artificial scape and contraction of the less artificial, the fringe is said to be advancing, and if vice versa, retreating. In some cases a static phase may develop and perpetuate the problems of the area indefinitely.

Fringe migration has an economic aspect. Retreating fringes are associated with depressed conditions and necessitate planning easement, as for example when an extractive land use declines and the population drifts away, transforming townscape into derelict rurban fringe. This is clearly a negative transformation of the land. Advancing fringes, on the other hand, are associated with economic growth, and may constitute positive transformations, as for example when agriculture reclaims wildscape or a town grows in response to a new industry. If, however, growth becomes too rapid it may cause inflation which is again negative because it gives rise to problems which call for planning action in the form of restrictive control.

The scape systems with their subsystems, and the fringe transformations with their phases and stages, are multiple concepts with functional, spatial and economic facets. Further facets such as location theory can also be traced in land uses and analysed at progressively more detailed levels. This complexity and interweaving of ideas in land-use studies is similar to that found in geography as a whole, and this is one reason why such studies are considered central to the subject. At the same time they remain the study of a single variable, the use made of the land surface. The combination of complexity and simplicity can sometimes allow land use to function as a substitute for the more elaborate web of constituent causes that are responsible for it. An example may be quoted from Canada.

Since much of Canadian agriculture is in a depressed state as a result of increased competition from better-favoured areas further south, the Canadian Agricultural and Rural Development Administration is at pains to identify the most marginal areas with a view to diversifying employment opportunities and consolidating the viable farms. To this end the Canada Land Inventory was set up to map a number of variables: soil capability classes, climate, unemployment, farm income levels, land use, etc., and to explore their interaction. A sample area about sixty miles west of Stellarton, Nova Scotia, may be discussed. The soil capability map shows non-agricultural soils of class 7 on the Cobequid Hills and fairly good arable soils of classes 2 and 3 on the coastal lowland to the north. These latter soils would not normally produce marginal conditions, but in this area their quality is offset by factors of severe east-coast climate, poor accessibility from Canada's populous heart, low levels of education, etc., which all have to be taken into account and given accurate weightings if the resultant of their interaction is to be correctly ascertained. All this can be avoided if the land-use map is used directly, since land use is itself the resultant of their actual, as compared to their calculated, influences. From the land-use map it is seen that the class 7 soils predictably form wildscape, while the class 2 and 3 soils, unpredictably, form a marginal fringe with only 27 per cent by area under crops and grass. Land-use analysis can thus provide a quick and sure way of identifying marginal areas by their spatial and quantitative patterns on the map.

Recreation affords another example of the employment of land-use territories to analyse a problem of topical interest. It has become a matter of growing concern as the result of today's increased leisure time and mobility. Recreational land use, like any other form of land use, may occur in any of the territories, and if tackled as a whole presents an amorphous and formidable problem. However, if broken down by territories it is seen to consist of five separate problems, each more manageable in itself. In the townscape there exists a large demand for recreation which has stimulated entrepreneurs to offer a wide range of entertainment facilities. Here the task left to local authorities is the provision of residual, non-profit-making facilities such as parks and open spaces. In the rurban fringe both the demand and the entrepreneurs are still close at hand, and because of lower land prices the latter are able to offer more land-extensive enterprises such as golf courses. The need here is to control their location in order to avoid fragmentation of farmland. In the farmscape a different problem arises. Strong recreational pressures may damage crops or fences, adding to the farmer's costs and detracting from his efficiency. These disadvantages can be outflanked by directing attention to specific recreational foci, such as stately homes or new country parks, or even beyond the farmscape altogether, into the marginal fringe. The marginal farmer is potentially more welcoming to the tourist, since he may wish to supplement his meagre farm income by providing accommodation or a caravan site. The difficulty here is his remoteness from the demand, and planning co-ordination is required to make the complementary needs of entrepreneurs and clients mutually known and actively provided for. Finally, in the wildscape, yet another aspect emerges. Here the recreational magnet is solitude and natural beauty, yet unless some kind of tourist provision is made beauty spots may tend to become open sewers or litter dumps. Facilities to offset this must be located unobtrusively, although not so unobtrusively as to be overlooked by their intended users. Frequently they must be the work of the local authority, since the wildscape itself has few or no residents.

So far this chapter has been concerned with man's basic space resource, land, and the chief inputs for its use which are energy and knowledge. But the industrial age has also seen a great burgeoning in the use of other space resources, which are now being consumed at a faster rate than ever before, and are also menacing land, water and air by the polluting effects of their waste products. Some of these resources, notably minerals, are not inexhaustible, and constitute a robber economy which was at one time a cause for alarm to geographers, since if even a few minerals should become unobtainable the lack of these could vitiate the use of others. This prospect of ultimate crisis is a cogent argument in favour of resource conservation, even though today the situation is viewed rather more optimistically in the light of the chemical revolution. The chemist has not only provided fertilizers, which have potentially removed the spectre of food shortages, but has also developed plastics, synthetic fibres and a wide range of other man-made materials. New materials, and old materials adapted to new uses, suggest even greater possibilities of resource substitution in the future. The biosphere, with its capacity for reproduction, may be called upon to furnish new synthetics industries with a new range of raw materials, and greater use may also be made of that scarcely tapped seven-tenths of the earth's surface, the sea. Ocean water and ocean floor may well become the mines of the twenty-first century, and if controlled marine biological production makes further advances the sea, like the land, may be progressively converted from wildscape to farmscape.

The optimistic view depends, however, on continued innovation by the human race to improve the quality of the time resource and, through it, the space resource. Such innovations are essential for continued economic growth, which has not yet achieved a sufficient rate in Britain despite a quarter of a century of government exhortation. In this context geographers may be interested in D. C. McClelland's *The Achieving Society* (1961), in which fourteen hypotheses of economic growth are tested against actual economic growth in forty countries. Nine are found to have no predictive value and the remaining five some value, but none of them are as closely correlated with growth as is an alternative hypothesis of dominant psychological motivation which recognizes three driving forces of achievement, friendship and power.

During the Elizabethan age of discovery, and again during the industrial revolution, the dominant type of motivation among Britons, as quantified from an examination of contemporary literature, was for achievement, and this produced positive results in the country's economy and civilization. In 1950, however, the dominant motivation was for friendship and not achievement; this would appear to provide some explanation for the post-war lack of impetus in economic growth. Currently there are certain indications of a trend towards power-seeking motivation. This features as one of the two main elements in the student unrest of the sixties, and if continued it does not presage a happy future. In the history of the world, motivation for power has frequently been associated with the subsequent decadence and decline of a civilization, unless, as happened in eighteenth-century Britain, a new wave of motivation for achievement sets in to reverse the trend.

Geographers are naturally interested in the spatial distribution of motivation for achievement, which varies very widely among the developing countries and allows us to venture the prediction that some, such as India, will achieve a substantial measure of development by the end of the century while others will make little progress. On a wave of achievement motivation all kinds of positive changes can be introduced in fields which seem to present insuperable difficulties during a non-achievement phase. Land ownership patterns may be cited as an example which are often believed to exert a stranglehold on progress. Large latifundia with absentee landlords, or small fragmented parcels with dense rural population, are both problem-ridden; a modern British example is the Lower Swansea Valley where the existence of thirty-seven different land owners was seen as a real obstacle to the reclamation of derelict land. Yet in a phase of achievement motivation land ownership problems are swept away without great difficulty, as witness the enclosure movement of the eighteenth and nineteenth centuries.

Achievement-motivated individuals and nations do not become enmeshed by their problems but focus their energies upon their opportunities. Geography today is opportunity-oriented, and it approaches problems with the positive aim of finding solutions for them. The latter approach is the province of applied geography. Land-use studies are not, *per se,* applied, but pure geography. They seek to discern order and pattern in existing reality rather than to create new order and pattern to adjust reality. But because they compare many existing patterns and assess their differential effectiveness, such studies can hardly avoid developing multiple links with applied geography, which is discussed in the next chapter.

FURTHER READING

ADDISON, H., *Land, Water and Food* (Chapman & Hall, 1961).
BEST, R. H. and COPPOCK, J. T., *The Changing Use of Land in Britain* (Faber & Faber, 1962).
BRAMLEY, M., *Farming and Food Supplies* (Allen & Unwin, 1965).
BROWN, H., *The Challenge of Man's Future* (Viking Press, 1958).
CIPOLLA, C. M., *The Economic History of World Population* (Pelican, 1964).
COLEMAN, ALICE, 'A Geographical Model for Land Use Analysis', *Geography*, Vol. 54, 1969.
HUNTINGTON, E., *Mainsprings of Civilisation* (Mentor Book, 1959).
JACOBS, J., *The Death and Life of Great American Cities* (Jonathan Cape, 1962).
STAMP, L. DUDLEY, *The Land of Britain: Its Use and Misuse* (Longmans Green, 1962).
STAMP, L. DUDLEY, *Our Developing World* (Faber & Faber, 1960).
SYMONS, L., *Agricultural Geography* (Bell, 1967).

10

Applied Geography

John W. House

In any discipline with a concern for mankind there is an urgent need for concentrated study of the many growing problems of economy and society. This study should not be solely academic in character but increasingly with a view to action. Action should be based on the co-ordinated analyses and advice of a variety of disciplines concerned with problems in common. Thus and only thus, to quote George Brown, can 'concentrated hard work and study by all result in sensible planning decisions'.

The advent of social science academics and those trained by them into the world of policy making, decision taking and decision testing is relatively recent but, once begun, has grown rapidly to meet the needs of the times. Geography became involved in this at an early date; its commitment has been progressive, and from this mounting preoccupation there has arisen one of the discipline's most intriguing growth foci, applied geography.

Science has long recognized the validity of applied aspects of academic disciplines, notably in mathematics and economics. The distinction of 'pure' and 'applied' branches has progressed sufficiently to give the latter a large measure of independence in teaching and research, with no lesser intellectual curiosity and respect. There has arisen a two-way interchange of ideas, techniques and results between the two branches within the framework of a coherent parent discipline, and this has been to their mutual advantage, affording greater depth of perspective.

Such is the currently developing position within geography,

and some of its most formative work is clearly focused in the applied field. There is some difference of opinion as to exactly how independent the applied branch should become. To some a separate identity for applied geography seems meaningless since most geography is capable of being applied, and much often is, and long has been. This is particularly true of human geography and also to some extent of physical geography, and much that geographers have written has been applied even though the original work was not carried out with the idea of application in mind. Other thinkers, however, take the opposite view, that the application would have been more effective if the relevant problem had been investigated with more vigorous definition and measurement, with a clearer relationship to the practical context, and above all with a view to specific recommendation and action. This would direct attention more sharply to precisely those problems where action is required and in respect of which novel methods and perspectives may have to be developed. It would encourage applied geography's distinctive research objectives and techniques, its special attitudes towards team work with members of other disciplines, and its characteristic ways of formulating and presenting results so that they embody recommendations. In this way applied geography remains within the widely accepted traditions of geography as a whole, but extends the frontiers of methodology, teaching and research as any newly formulated systematic branch is likely to do.

The range of applied geography

The term 'applied geography' does not always mean the same thing. It carries different connotations in different lands and under different political dogmas and ideologies. In the socialist lands of the U.S.S.R. and Eastern Europe, and indeed to Marxists everywhere, applied geography is seen as central to the discipline where it supports the struggle of socialist man to impose his will on the environment. Consequently much of the undergraduate training is problem- and action-oriented, and both physical and human researches are more sharply polarized on the requirements of the community and the state. This is a far cry from the free enterprise standpoint of the U.S.A. where a leading geographer has recently asserted: 'Applied geography is business geography.' While this is an underestimate of the increasing contribution that American geographers have been making in public agencies, it carries the interesting implication that the applied geographer is a practitioner soon to be capable of living by his expertise in the business world.

The British perspective on applied geography has always involved a practical and empirical approach, starting from foundation studies in several fields where problems were seen to be acute. Social and economic problems were studied to some extent from the late nineteenth century, but the major development of action-oriented research came with the severe economic depression of the 1930s, in work such as that of G. H. J. Daysh on distressed areas, L. Dudley Stamp on land utilization, E. G. R. Taylor on the distribution of industrial population, and later, A. E. Smailes and R. E. Dickinson on urban geography in relation to planning. The same problems that stimulated the growth of applied geography often also called for action by the governments of the day at central, regional and local level. Thus the needs of the time drew geographers into advisory and consultative work, for which they had to enlarge and refine the fields of application of their subject. This was preparatory to their being recognized as full professionals and appointed to specialist posts in most of the organizations which had responsibility for the burgeoning branches of planning.

The Second World War gave a sharp impetus to the involvement of geographers in both civil and military planning. The formation of the Ministry of Town and Country Planning in 1943, followed by the comprehensive planning legislation of the 1947 Act, introduced geographers as acknowledged specialists both in central government research organizations and also in the statutory planning authorities of the counties and county boroughs. These new opportunities have been perceptibly and progressively capitalized, and a great deal of work has been done to adapt the geography of teaching and research institutions to 'real-life' situations.

As a result, the professional employment of geographers in planning has come to be the major single career opportunity after education. In addition to government posts, British geographers have been appointed to voluntary bodies concerned with planning and have also occasionally acted as consultants in research teams or on a private basis.

Even so, many geographers feel that the role of geography in planning could be significantly extended, and if this is to be so it will surely require more rigorous training in applied geography. There is also a need for more co-ordination between government research organizations and university departments to assess what geographers individually and collectively have achieved in the applied field so far, and to consider how agreed objectives, methods and working programmes could be of advantage to both sides.

Applied geography today

Applied geography plays three simultaneous roles. In the first place it is no more than a logical and natural development of 'pure' geography using orthodox concepts and techniques. Second, it is a distinctive subdivision of geography with its own specialized concepts and techniques. And third, it is a frontier discipline linking geography proper with planning proper.

In its first role, applied geography emphasizes the conventional focus upon the spatial variable in habitat characteristics and socio-economic conditions and problems. Its methods show the same pre-occupation with orthodox factorial analysis by sieve techniques and cross correlations between physical and human factors, and they also include the fashionable and rapidly growing quantitative methods of analysis. The detection, definition and dissection of significant areas fulfils the same central function as in pure geography, and cartography is also used in a thoroughly conventional way both as a tool of analysis and as a method of presenting results.

In its second role there appear differences which distinguish applied from pure geography. The fact that applied geography is action-oriented and devoted to serving the needs of a definite client often determines the kind of problem tackled and the way in which a solution is attempted. Clients are most likely to be planners or public bodies; private corporations or individuals figure to a lesser extent. Their needs may be extremely diverse but this does not mean a piecemeal collection of knowledge. Applied geography must develop as a coherent and systematized body of techniques and concepts if it is to meet the varied demands.

Whereas most conventional geography is concerned with the present as it has emerged from the past, in applied geography a concern for the future is cardinal. The time span is the proximate and middle future, covering about fifteen years, which is also the usual time perspective of most planners. This goes somewhat beyond the limits of straightforward programming by projection of existing trends; it may involve the initiation and location of new projects, in fact the foundations and essence of a strategy. Longer-term planning objectives for around the year A.D. 2000 are also within the scope of the applied geographer, although these are clearly more speculative.

Even qualified prediction necessitates the greatest precision that the data will bear. Long-term forecasting may involve mostly a general conceptual description within a loose framework, but in shorter-term work quantitative techniques must play a decisive role. Systems analysis, the building and testing of predictive models, the use of computers, the creation of data banks and data retrieval mechanisms are all essential. Some have been pioneered in other social sciences and transplanted to geography, others have been developed within geography itself. They are not unique to applied geography. Nevertheless, because applied geography has to measure in order to solve and convince, and because it must be thoroughly inter-disciplinary within the planning field, it needs the numerate approach even more than most branches of the subject.

At the same time it is recognized that quantification of data and sophisticated processing do not provide optimum solutions automatically, nor even necessarily indicate a clear priority among alternative recommendations. Some variables are difficult and even impossible to quantify. In problems of land-use conflict the most valid method is generally the cost-benefit study, but it is not uncommon for residual value-judgements to be required after the computations have been completed. At this point the applied geographer, the user and the planner are almost indistinguishable, though it is the latter two alone who go on to executive action.

Because of its preoccupation with change, trends, possible changes and their programming, applied geography places great emphasis on the dynamics of situations. Most of its work falls within the following well-defined fields: the problems of using natural resources in the contexts of profitability, welfare or the legislative framework; the needs of communities for recreational space, national parks, holiday areas, mobility, etc.; the spatial ordering of economy and society including the administrative-functional pattern; the dynamics of city growth and change, urban locational systems, renewal, transportation, congestion, zoning, overspill, new towns and green belts; the processes of regional development, growth, decay and imbalance. In addition the needs of private enterprise are also being served; recently active fields of study include supermarket chains, industrial location and transportation media.

The third role, the special relationship with the field of planning, is borne out by the fact that 10 per cent of men and 7 per cent of women geography graduates went directly into central or local planning in one recent year in Britain. Furthermore, no less than 30 per cent of the candidates for associate membership of the Town Planning Institute recently had a first degree in geography.

Let there be no mistake, however, the applied geographer is not a substitute for the planner. On the other hand, it is no longer the practice to restrict geographers to the survey stage of planning and to exclude them from the analysis and plan stages. It is now increasingly realized that planning is a continuum and a team activity, in which the composition of generalists and specialists is best left flexible, to be determined according to circumstances, problem and place. No one specialist discipline has the automatic right to lead the team regardless of circumstances.

The tendency towards group activity is powerfully reinforced by the fact that when social scientists focus upon common problems their different perspectives begin to converge. Techniques are often developed and used in common, although the central objective of each discipline retains its distinctive identity.

These days it is hard to identify what is unique to a discipline in the planning field, but geography's contribution to the common weal lies in its emphasis on areal variation, its distinctive forms of factorial analysis, its regionalization procedures, its comparative study of case histories and its extensive use of cartographic techniques. Of course, as specialists become more deeply involved in the planning process, they tend to merge identities, and as they increasingly acquire the complementary training of the planner, they pass into that profession. Even so, their early training in a specific discipline carries over, like the 'smile of the Cheshire cat', to influence the kind of thinking framework that they continue to use, and the geographical influence has been among the most important in post-war years.

In Britain since 1947 there has been some divorce between physical planning and social and economic planning. The former is the statutory obligation of the counties and county boroughs to prepare development plans and periodic amendments for ministerial approval, while the latter has generally been formulated at national level as the concern of several, sometimes loosely co-ordinated, ministerial programmes. Thus local authority planning departments and central government ministries have offered two rather different types of employment for geographers. The intermediate regional level of planning, which would be even better suited to geographical talents, has until recently unfortunately been conspicuous by its absence in Britain. Only in 1965 did it tentatively emerge in the form of regional economic planning councils, which potentially present the geographer with better opportunities to ply his craft.

Geographers in planning

Most geographers in central government planning are located either at the London or regional offices of the Ministry of Housing and Local Government, or in the Scottish Development Department. They work on physical, social, economic and demographic aspects of urban and regional planning and on the review procedure for local authority development plans, as well as advising in other fields such as housing and local government, new towns, the hierarchy and functions of service centres, etc. The maps section of the Ministry has a general service function in cartography, and has also produced the ten-mile maps of Britain and the national desk atlas.

In Scotland there has been greater and earlier emphasis on regional planning, allowing geographers to exercise a wider range of their skills, including regional surveys, rural studies, migration flow analysis, growth point studies in the Central Lowlands and the compilation of industrial site dossiers. Some preference is indicated for geographers with research or practical experience in urban studies or sociology applied to planning.

Geographers have already been appointed at the policy formulating level of four of the new regional economic planning councils, and the research support teams include geographers from the Department of Economic Affairs and other ministries as well as the Ministry of Housing and Local Government.

At county level the work of geographers is less akin to their conventional academic studies, but merges more rapidly and integrally into professional planning expertise. To equip themselves for this work, and also to ensure transferability in the career structure, a planning qualification is usually sought soon after completion of the first degree, either in the form of a diploma or through the Town Planning Institute examinations.

Geographers are also appointed as specialists, but the numbers vary markedly according to the local authority. One major northern planning department recently had a roll of sixteen geographers: two in senior administrative posts;

eight in the urban plans section working on land-use plans and town maps; three in the county map, policy and research section of the development plan organization; two in development control and one working on critical path analysis for environmental improvement. The majority had become, or were in process of becoming, fully professional planners. Other authorities normally employ fewer geographers but in a similar range of roles. Gone are the days when the geographer was limited to the survey stage.

Local government reform is very much in the air at the moment with the likelihood of larger units that are better related to the social and economic realities of contemporary Britian. The planning legislation of 1968 breaks away from the rigidities of detailed land-use programming in favour of more general strategic plans seen as emerging from the framework of economic planning regions. These will lead to the formulation of local plans for districts and action areas, and citizen consultation is envisaged as playing a bigger part than hitherto. There is prospectively greater opportunity both for specialists and for team work, while the citizen-geographer could be a valuable element of the more informed electorate needed for increased public participation. This strengthens the educational argument for the advancement of applied geography to bridge the gap between geography and planning.

Turning to the topic of private consultancy, the applied geographer can rarely be so bold as to set up a brass plate and live by his expertise. Such good fortune is enjoyed by but a few in the United States, mainly in the fields of store or plant location and central business district research. Consultancy in Britain is usually through private research contract or participation in an interdisciplinary group of consultants. These forms originated during the late war and early post-war periods of consultants development plans for the Clyde Valley, Middlesbrough, the West Midlands, Greater London and south-east Scotland. The results derived from geographers' participation were fed back into the developing stream of applied geography, providing major contributions in settlement and social geography especially.

Targets and achievements in Britain

The applications of geography are potentially as wide as the subject itself. British geographers have already made contributions in the fields of coastal erosion, dune stabilization, flood control, water supply, long-range weather forecasting, atmospheric pollution, soil survey, agricultural land classification, etc. Nevertheless, the applied scene in Britain is heavily dominated by planning because of the urgent need for attention to problems of growth, stabilization and, at times, phased decline in resource use, settlements, industries and regions.

Within planning there are certain key fields of central geographical interest, representing targets and reflecting achievements. The outstanding topic at present is probably the identification and structural analysis of all types of human locational systems, especially urban systems. This demands research at two ends of a spectrum. At the general end there is the formulation of models of systems, and at the particular end there is the continuing development of individual case studies, pursued at depth and on a comparative basis. Systems of cities, economic subregions and socio-economic community structures are usually exceedingly complex and often open-ended. Research makes the attempt to identify links and bonds, flows and forces, in order to understand operative mechanisms as a basis for modifying or manipulating them through the planning process.

The land-use-transportation studies that are presently in progress in the major British conurbations will provide a key to the prediction and programming of desirable land-use patterns in the regional geography of the future. Although this aspect of urban research is largely the preserve of consultant planning teams, it articulates clearly into conventional urban geography and is a logical outcome of systems-based research. In order to extend the concepts of the hierarchy of places and locations, geographers are active in formulating predictive models, establishing data banks and investigating information retrieval.

Regionalization in all its forms is as much a spearhead in

applied as in pure geography, but here its purpose is more defined. Regions are studied not simply as the spatial reflections of distinctive man-environment relationships, but with a view to the more effective ordering of economy and society. Another application is the study of the relationships between the varied regional patterns that have developed for different purposes in the same area; this assists the devising of a hierarchy of local government units that will be related to pre-existing patterns of economic regions or social groupings. Other potent subjects for research are networks or urban centres of varied size, rank and function, and the regional effect of new towns, industries, hospitals or supermarkets, etc.

The regional impact of government's locational policies has also been the subject of a series of studies over the past thirty years. Researches into imbalance of prosperity, growth industries or social provision as between the regions, have been complemented by studies of particular problem areas which are often marginal urban, industrial or rural districts. Geography has also contributed with the other social sciences to investigations of the ingredients of regional development. One aspect, which has not yet been studied to the full, is that of the effect of particular transportation projects, such as the motorway programme, the contraction of the rail net, the building of the Severn Bridge, Teesport, the third London airport proposals, and so on. Another aspect of regional development which is increasingly attracting geographers' attention is the key role of the service industries, a breakaway from the notion that only tangible resources such as land and the products of field, mine and factory merit geographical study.

Indeed the general point can be made that man himself is becoming a more important focus of research in both pure and applied geography. Studies of changing demographic structures and patterns are being complemented by work on that most elusive yet most central phenomenon in an increasingly mobile society—the migrant, his household and his family. Geographical advice has been sought and tendered on the nature and significance of interregional migration movements in Britain, and it has proved possible to interpret the complex motivation responsible for them.

Land-use conflict in modern society is a further central theme that is as cardinal in geographical research as regions and people. Such conflicts may exist even in the most remote areas where farming of several types may contend with forestry, reservoir development, quarrying or recreation, but they increase in frequency and complexity in more populated areas. Conflict is characteristic in the central areas of cities, at the core and perimeter of the central business district, and also along the significant line or zone where town merges into country. The resolution of land-use conflict in the light of priorities determined by the common interest is the planner's most crucial task and one to which geographers can contribute. Indeed the problems of conflict are essentially geographical right up to the point at which analysis passes into policy making. The identification, classification and mapping of land uses in town and countryside has been a traditional geographical activity for forty years, but much remains to be done to harmonize and improve the classifications used by geographers and other land-use analysts before all can benefit from the establishment of regional data banks for the storage and processing of the agreed data.

From land use to the assessment of land potential is a logical but difficult step needed as a prelude to interpreting pressures for land-use change with their variable incidence and rates of progress. Such interpretations have to be based not only on studies in space and time, but equally on the dynamics of change itself. This is the central concept of applied geography.

The field of study of the interpretation of land-use change is exceedingly wide. One aspect of recent interest has been the growing pressure for recreational use, which involves a growing need to reconcile multiple claims on limited territory. This has led to special studies of the relevant service industries, such as the tourist industry, which was little considered in geography until recent years.

Closer to the engineering profession and landscape architecture have been studies of sand and gravel deposits; of the physical and aesthetic merits and disadvantages of alternative courses of motorways; of the need to preserve coastlines of scenic beauty; and of the restoration and rehabilitation of landscapes, such as the Lower Swansea Valley which has been the subject of a multidisciplinary research project.

This thumb-nail sketch of the wide diversity of significant problem-oriented fields of research in planning could be accompanied by a catalogue of geographical contributions in virtually every case. Targets and achievements taken together indicate the validity of applying geographical techniques to real-life problems. Geography must not, however, rest on its laurels or lapse into the belief that, because so much of it is potentially useful, there is not a continuing need to meet the requirements of others still more closely.

A world perspective

Applied geography came of age in the international geographical community at the London meeting of the International Geographical Union in 1964. A commission on world applied geography was established then, and at the 1968 meeting in New Delhi it was confirmed as one of the only two standing (or permanent) commissions of the I.G.U., alongside that for national atlases.

The commission first attempted a world survey of the tasks and achievements of applied geographers, and this revealed the universality of the trend towards a problem- and action-orientation in geography. It also brought to light the extreme diversity of the forms that application was taking and the wide variety of types of user, and it established in prospect a great corpus of knowledge and experience in applied work throughout the capitalist, socialist and developing worlds.

The 1968 New Delhi meeting defined the main tasks facing applied geography. The first is the need to share and refine research methods and to work towards spatial inventories of all types of phenomenon, with improvement of data storage and retrieval systems. Only thus can the infinite detail of natural and human features be reduced to order and made capable of scientific interpretation.

Second, geographers have to make mental adjustments if they are to move more effectively into the applied field. It requires a major effort to rethink what is already patently useful, in ways that will be meaningful to others and directed towards the elucidation and solving of problems. This was widely recognized as meaning that there should be complementary training prgrammes in applied geography alongside the main geographical curricula of the world's universities and institutes. Such facilities are in fact increasingly being provided. Twenty British universities now list undergraduate courses in applied geography, and in four cases these are compulsory, while there are at least two postgraduate courses—at the London School of Economics and the University of Newcastle. In some countries, notably those of the socialist world, the production of geographer-specialists for careers in planning organizations and government service starts early and is substantial. The geographers thus trained tend to work with distinctive sets of tasks, in which the effective use and co-ordination of natural resources, and the economic regionalization programmes of governments are central themes. In the developing countries, not surprisingly, a different picture emerges. Because of the general scarcity of further education there are as yet few applied geographers and little undergraduate instruction oriented to application. Since these areas also face the most severe long-term problems, they will be the scene of the greatest challenge to established applied geographers during the next few decades.

This chapter has shown that geographers are increasingly regarding themselves as men of action, and that this involves a logical, purposeful extension of the conventional boundaries of the subject without, however, changing its essential nature. The manifold problems of the times dictate that 'he who will not advise, consents', and, in the words of the late Sir Dudley Stamp, 'the applications of geography one seeks are for the benefit of society'. It is hoped that all intending students of geography will understand that their subject has a genuine concern for the problems of peoples and places, and a growing effective contribution to their understanding and solution. In an age of rising social commitment geography is a subject eminently worth studying.

FURTHER READING

ARVILL, R., *Man and Environment* (Pelican, 1967).
DAYSH, G. H. J. (ed.), *A Survey of Whitby and the Surrounding Area* (Shakespeare Head, 1958).
FREEMAN, T. W., *Geography and Regional Administration* (Hutchinson, 1968).
HALL, P., *London 2000* (Faber & Faber, 1963).
HILTON, K. J. (ed.), *The Lower Swansea Valley Project* (Longmans Green, 1967).
HOUSE, J. W. and FULLERTON, B., *Tees-side at Mid-Century* (Macmillan, 1960).
JACKSON, J. N., *Surveys for Town and Country Planning* (Hutchinson, 1963).
OXENHAM, R. J., *Reclaiming Derelict Land* (Faber & Faber, 1966).
STAMP, L. DUDLEY, *Applied Geography* (Pelican, 1960).
STARR, R., *Urban Choices, the City and its Critics* (Penguin, 1966).
Land Use and Resources: Studies in Applied Geography, a memorial volume to Sir Dudley Stamp (Institute of British Geographers, 1968).

11

Geography in British Universities

Robert W. Steel

Those who fill up the forms produced by U.C.C.A.—the Universities Central Council on Admissions—and who read the accompanying booklet find themselves faced with information supplied by nearly forty departments of geography (and related subjects such as environmental sciences) and with a remarkable range of about seventy degree courses. They may be surprised to learn that not so many years ago the problem facing the would-be student of geography was not to choose between a galaxy of courses and departments but to discover the whereabouts of the handful of departments in universities and colleges where geography was in fact recognized and could be studied. For though geography is one of the oldest of subjects, going back at least to classical times, as shown by the writings of men like Strabo and Herodotus, it is relatively young as a university discipline and most of its growth in universities has taken place during the present century, and particularly in the years since the end of the Second World War. This largely explains why relatively few of the graduates of the past read geography even as one subject in a general degree, let alone as a single honours subject. Indeed, until recently, many of the senior geographers teaching in university departments of geography were the products of other disciplines, and there was no professor of geography holding an honours degree in geography until shortly before the end of the Second World War.

Many different kinds of people have contributed to the geographical knowledge and understanding of the world, and many of them would not have regarded themselves as geographers. They have included travellers, explorers, soldiers, statesmen, mathematicians and philosophers; and anyone embarking on courses in geography should read about them in books such as G. R. Crone's *Background to Geography* (1964), so that he can appreciate the long tradition of the subject and at the same time see how men have tried to describe and analyse and understand the features of the earth's surface, whether natural or man-made. Such a study of the history of geography would include a wide variety of interesting matters relating to individual geographers and other persons. Thus in the year 1187 a Welsh scholar in Oxford, Giraldus Cambrensis, is said to have read aloud his *Topography of Ireland* for three whole days. In 1574 another Oxford scholar, Richard Hakluyt, a student of Christ Church, gave lectures on geography and later produced a series of volumes called *The principall navigations, voyages and discoveries of the English nation:* these supplied his contemporaries with full accounts of all the discoveries of the busy period in the latter half of the sixteenth century following on the Great Age of Discovery. A century later Isaac Newton, the discoverer of the law of gravitation, edited a translation of a book on *General Geography* first published in Amsterdam in 1650 by Bernhard Varenius. Newton believed that the work would be of benefit to the students of Cambridge. During the eighteenth century, Captain James Cook made important contributions to the geographical knowledge of the world through his extensive voyages in the Pacific Ocean, and in 1788 a dining club called the African Association was founded to encourage the discovery and opening up of Africa. In 1931 this was absorbed into the newly created Royal Geographical Society. One of the Society's first secretaries, Captain A. Maconochie, gave lectures in geography at University College, London where from 1833 to 1836 he was designated Professor of Geography —the first to hold such a title in the United Kingdom. King's College London appointed its first Professor of Geography, William Hughes, in 1863.

It is, however, the latter part of the nineteenth century that is really significant for the establishment of geography as a discipline in British universities. Two German scholars, Alexander von Humboldt and Carl Ritter, are sometimes referred to as 'the fathers of modern geography'. They both died in 1859, the year in which Charles Darwin published his *Origin of Species,* which had such a profound influence upon scientific thought. Following them other Germans like Friedrich Ratzel developed studies in 'anthropogeography' or human geography, while in France much of the sociological work of men like Frédéric le Play had considerable geographical content. In Britain the Royal Geographical Society commissioned J. Scott Keltie to produce a report on geographical education which was published in 1885, and two years later Halford J. Mackinder read a paper on 'The New Geography' at a meeting of the Society. This lecture, together with the Scott Keltie report, resulted in the Society agreeing to provide some financial support for the establishment of the subject at Oxford and at Cambridge, at that time almost the only universities of which the Society took any note. Mackinder himself was appointed Reader in Geography at Oxford in 1887, while at Cambridge F. H. H. Guillemard and F. Yule Oldham were appointed lecturers in the following year: the latter, with the title of reader, remained in charge of the subject at Cambridge until 1908.

In universities and colleges elsewhere, various geography courses were already being given. In Liverpool, for example, there may have been a course on geography in the session 1886-7, and certainly regular lectures on 'commercial geography' were delivered from 1888 onwards. At Manchester, geography courses date from 1892. The first School of Geography appears to have been that at Oxford, which was established in 1899 with a staff of three.

The first professor of geography in modern times, L. W. Lyde, was appointed to a Chair of Economic Geography at University College London in 1903 : he had been trained as a classicist, a background that is clearly revealed in some of his writings such as *Peninsular Europe* (1931) and *The Continent of Asia* (1933). P. M. Roxby, straight from the Honour School of Modern History at Oxford, where he had also attended geography lectures by Mackinder and others in the School of Geography, was appointed in 1905 to give lectures in geography in the Department of Economics at Liverpool. H. J. Fleure, a zoologist by training, went to Aberystwyth in 1907 to take charge of the development of geography in the University College of Wales. Mackinder left Oxford in 1905 after becoming Director of the London School of Economics. He was succeeded by A. J. Herbertson, who had studied, among other subjects, botany and meteorology at Edinburgh and who had the distinction of holding a personal chair in Oxford from 1910 until his death in 1915.

Throughout the history of mankind, war has seemed to stimulate the study of geographical matters, and during the First World War several geographers were engaged in preparing a series of geographical handbooks for the Naval Staff Intelligence Department. Other people had their interest in the geographical background to international affairs aroused by service in the forces and, in some instances, by later involvement in the work of those responsible for drawing up the peace treaties after 1918. In the universities, despite the very considerable disruption of the war, there were some very important developments. In 1917, a chair of geography at Liverpool was endowed by John Rankin, and the University simultaneously established an Honours School of Geography in the Faculty of Arts, the first in any British university. Very soon afterwards Aberystwyth established the Gregynog Chair of Geography and Anthropology with honours schools in the Faculties of Arts and Science. Some of the London colleges also created honours B.A. and B.Sc. courses. These courses were additional to the then existing B.Sc. (Econ.) courses which had been given in the London School of Economics from 1900 and in University College from 1912. One of the first of the honours geography graduates of the University of London was L. Dudley Stamp. The new standing of the subject in London was recognized by the appointment in 1922 of Sir Halford Mackinder and J. F. Unstead to chairs of geography at the School of Economics and Birkbeck College respectively.

In the next few years there was steady development of the subject in many universities, and those in charge of departments of geography were sometimes promoted to readerships and professorships. During the 1920s, chairs were created in the University College of the South West (now the University of Exeter), Manchester, Sheffield and Edinburgh; but it was only in 1931 and 1932 that Cambridge and Oxford respectively—the two universities helped by the Royal Geographical Society more than a generation previously—established chairs. The only other chair founded before the outbreak of the Second World War was at Bristol where the lecturer appointed in 1926 became a professor in 1933. The 1930s were very difficult years in all British universities, financially and in other ways, and geography, which still had to establish itself as an academic discipline in the eyes of many, was especially affected by the shortage of money and the general lack of university expansion.

Nevertheless, increasingly large numbers of university students took degrees in geography between the wars, and a high proportion of them became teachers, chiefly in secondary schools, since at the time there were few outlets in other branches of education. Very few of these graduates were able to continue their studies with full-time research, although many undertook part-time research along with their paid employment. A particularly valuable contribution made by many of these research workers was the preparation of some of the county memoirs that formed an indispensable part of the First Land Utilisation Survey of Britain organized from 1930 onwards by Dudley Stamp at the London School of Economics.

This research activity of geographers, combined with increasing involvement in the study of the economic and other problems of many parts of Britain, especially around their own universities, helped to establish the nature of the contribution that geographers could make to the then young, and by no means widely accepted, field of town and country planning. Through the Royal Geographical Society a group of geographers under the guidance of Professor Eva G. R. Taylor produced a memorandum and a series of maps for the consideration of the Royal Commission on 'The

Distribution of the Industrial Population', which produced the Barlow Report in 1940 (Cmd. 6153). During the war several geographers were recruited to prepare for post-war planning and served in the Ministry of Town and Country Planning established in 1943. The report of the Scott Committee on 'Land Utilization in Rural Areas' appeared in 1942 (Cmd. 6378) and underlined the importance of a geographical understanding of land-use problems: Dudley Stamp, who had been the committee's vice-chairman, was appointed Chief Adviser on Rural Land Use to the Ministry of Agriculture. Other geographers served in the armed forces, where some were able to make use of their training in intelligence branches, in meteorology and climatology, and in surveying, while a large group formed the nucleus of the geographical handbooks section of the Naval Intelligence Division of the Admiralty, which produced a much larger and more elaborate—and geographically superior—series of handbooks than the comparable but smaller team of geographers had done during the First World War.

After the end of the war there was a tremendous flood of returned soldiers to the universities and a very remarkable increase in the number of students wishing to read geography. Some had started to study other subjects before, or early in, the war, and now wished to change their courses for a variety of reasons: many others were fired by the wish to know more about different parts of the world of which they had seen something while serving in the forces. The universities recognized the demand and where possible provided the teachers and facilities for newer subjects alongside more traditional fields of study. As a result there was a general expansion in the size and range of geography departments nearly everywhere, and this growth has continued throughout the post-war period, at least until the current financial stringency forced universities to look very cautiously at new growth, particularly where expensive equipment was involved. Early in the 1960s the Robbins Report on Higher Education (Cmd. 2154, 1963) encouraged further expansion. The number of university institutions recognized and financed by the University Grants Committee increased from about thirty to well over forty, and the entrance prospects for potential students improved accordingly, though not always for the subjects that they particularly wished to study.

The expansion in geography was concentrated largely in existing departments, and the annual number of honours graduates in geography increased from 700 in the mid 1950s to more than 1300 in the late 1960s. This number may well grow still larger during the 1970s, for there is as yet little sign of saturation point being reached in the employment of geographers, especially with the significant increase in the variety of outlets for graduates in the subject (see Chapter 12).

Today there are departments of geography in nearly every university. All those established by the end of the Second World War have chairs of geography, except the University of St Andrews. In the University of Wales the teaching of geography is concentrated in the University College of Wales at Aberystwyth and in University College, Swansea, and is not at present undertaken in Cardiff or at Bangor: a new department is planned for 1971 at St David's College, Lampeter. There are chairs in all the colleges that used to be associated with the University of London and now have independent university status, at Exeter, Hull, Leicester, Nottingham, Reading and Southampton. Of the universities built since the war, geography was one of the foundation chairs in the University of Keele, and the University of Sussex has a geographical laboratory, under a professor of geography, associated with the different schools of studies, including European, African and Asian studies. Strong departments of geography exist in some of the Colleges of Advanced Technology that now have full university status, as at Salford, where geography is closely associated with economics, and at Strathclyde, where it plays an important part in the degree courses in Liberal Studies. Environmental Sciences, with varying shades of meaning and with the role of geography differing considerably from university to university, are important schools in the University of Lancaster, the New University of Ulster at Coleraine, and the University of East Anglia (where the dean of the School of Environmental Sciences is also Professor of

L

Geography). In the University of Surrey there is an interesting and specialized development in the Department of Linguistics which has a Reader in Geography. Geography is also being developed rapidly in many colleges of technology and other similar institutions, some of which have very large departments of geography, as at Portsmouth and Cambridge, and it is a subject recognized to be of degree status by the Council for National Academic Awards.

Any attempt to tabulate all the courses now available in geography would be very difficult and perhaps misleading, and might not be very helpful to the would-be student, although such an analysis is done regularly by bodies such as the Careers Research and Advisory Centre (C.R.A.C., Bateman Street, Cambridge) and the Advisory Centre for Education (57 Russell Street, Cambridge). But with so many large departments it is hardly possible in these days to state that certain universities are particularly good or peculiarly suitable for the needs of an individual with a special interest. In earlier days distinctive traditions were often established, some of which have been maintained, though perhaps in a modified form. Cambridge, for example, was especially renowned for its physical geography; the first professor, F. Debenham, had been trained as a geologist and was a member of Scott's Antarctic expedition of 1910-13. At Aberystwyth geography was for many years associated in a joint department with anthropology. At Liverpool, where the first teaching was carried out in the Department of Economics, P. M. Roxby played a leading part in the development of human geography and had a special interest in countries overseas, notably in the Far East. Oxford, where so many of the early teachers of geography were trained at summer schools and in diploma courses long before the Honour School of Geography was instituted, was always regarded as having a particular interest in the study of regional geography.

Today, however, the situation is vastly different for, with more departments, and particularly larger departments, the range of teaching can be much greater; indeed, in most departments both undergraduate teaching and research are commonly undertaken in many fields over a broad front.

Thus Cambridge, despite its tradition in physical geography, has produced in the post-war years a group of outstanding social and economic geographers, several of whom now occupy chairs and readerships in other universities. The department at Aberystwyth is no longer concerned with anthropology as well as geography, and it now has considerable interest in physiography. Today any student graduating in a university department of geography is likely to be well grounded in the basic principles of both physical and human geography, and usually regional geography as well, though naturally some will have been more stimulated in certain fields than in others through the enthusiasm of individual lecturers who have taught them.

All departments of geography can, therefore, teach the essential elements of the subject, thanks to the availability of more staff than in the past. Larger staffs also permit more specialization than previously, and in some of the larger departments there is an almost overwhelming range of optional courses, usually at final-year level. There are also in some departments what are generally listed in the tabulations as 'compulsory courses'; everyone is expected to follow, say, a particular regional course or to learn a technique such as surveying. It must not be assumed, though, that such a department necessarily regards that region or technique as its own particular field of specialization; it just looks on these courses as especially valuable for all its students, without exception. Often the outstanding research interests of a department do not figure very noticeably in the undergraduate syllabus, but are concentrated at the postgraduate level where many students may be working in the department's special field of interest, whether systematic or regional. But there is usually a very close link between undergraduate and postgraduate work and the research interests of staff members. All teachers in universities—in contrast to those in some other institutions of higher education—are, in the terms of their appointment, required 'to teach and to undertake research', and the majority see teaching and research as equally important sides of their work. There is no doubt that teachers are nearly always most effective when lecturing about the matters that interest and stimulate them most—their own research work and the background to it. Perhaps it is not without significance that some of the departments with the largest undergraduate intake are also those with the best-developed graduate schools and the staffs that are most productive of scholarly writings.

To describe the special teaching facilities or the major research interests of individual departments, especially the larger ones, presents many difficulties today. Sheffield, for example, has a professorial interest in physical geography, yet from that department there have come in recent years important contributions to our understanding of the Domesday geography of northern England and to the geographical analysis of the 1961 census of England and Wales. In the same way there is no neglect of physical geography in, say, the Universities of Edinburgh or Newcastle upon Tyne, each of which has professors whose major research contributions have been in the fields of human and regional geography. It is well known that there is specialization in various systematic fields of geography in some universities and colleges—geomorphology, for example, in Birmingham, biogeography at Bedford College, London, or in the New University of Ulster economic geography at the London School of Economics, urban geography at Queen Mary College, London, and social geography at Liverpool—but it would be hard to produce either a complete or an up-to-date list. For regional studies there are perhaps fewer difficulties, though departments and schools of geography assess the importance of regional work in widely differing ways. One or two claim to do no regional geography, though one such university, Bristol, now has a professor of urban and regional geography; others do a great deal. Elsewhere the emphasis varies considerably between these extremes, and there is much variation in the attitude to regional work of those responsible for the teaching. Some departments expect every specialist teacher to have a regional interest as well; others would not want anyone to lecture on a particular part of the world unless he had at least some first-hand knowledge of that area and some personal involvement in it and concern for it.

Most departments base much of their work on the British Isles, usually with particular reference to the region in which they are situated, and departments in Wales, Scotland and Ireland, very understandably and reasonably, devote considerable time to the study of the problems of their own countries. Beyond the British Isles the continent of Europe generally, though not always, figures prominently in the regional syllabus. Despite the language problem in both teaching and travel, students have good opportunities to obtain some first-hand knowledge of Europe, and as the links between Britain and the countries of both the European Economic Community and the European Free Trade Association become closer, it is clearly desirable that young graduates should be as knowledgeable as possible about their neighbours on the Continent.

Different problems arise with English-speaking but more remote areas such as the U.S.A., Canada, Australia and New Zealand. Much-used books on North America have been written by senior members of the staffs at St Andrews, Edinburgh, and University College London, and North America has been visited by many British geographers in post-war years so that courses on the continent are fairly common. Australasia is also known at first hand by more geographers than used to be the case, and regional courses based on their experience are given in a growing number of departments.

In recent years two special committees of the University Grants Committee were set up to consider ways and means of improving the knowledge of overseas areas that had hitherto been inadequately studied in British universities. The Hayter Committee dealt with Oriental and African language areas in Asia and Africa, and the Parry Committee was concerned with Latin-American studies. As a result of their recommendations, inter-disciplinary centres of studies have been established in various universities. These include the following: Chinese (Leeds), Japanese (Sheffield), South-East Asian (Hull), South Asian (Cambridge), Middle Eastern (Durham), and East European and Russian (Swansea), in addition to the concentration of activity in these fields in the School of Oriental and African Studies in the

University of London. African studies are more widely spread over several universities with Hayter centres in Birmingham (West African) and Edinburgh (African). Geographers are associated with all these centres and in some cases play a particularly important part. Thus the directors of the South Asian Centre in Cambridge and of the Middle East Centre at Durham are geographers, and the editor of *Modern Asian Studies* is Professor of Geography at the School of Oriental and African Studies in London. Geographers with first-hand experience of teaching and research in various parts of Africa are to be found in many other universities. The Department of Geography at Liverpool, for example, has seven members of its staff with African experience, and the total number of geographers interested in Africa in the University of London's Departments of Geography is quite considerable. There are other groups of some strength in Swansea, Belfast, Salford and elsewhere.

The Parry Committee recommended five centres for Latin-American studies in London, Oxford and Cambridge and in the universities of the two great ports that have always had close commercial and other ties with Latin-American countries—Liverpool and Glasgow. All these centres have appointed geographers to their staffs who work in close association with the appropriate departments in the university; there are also geographers who have worked and travelled in Latin America in other universities including Leeds, Leicester, Manchester and the University College of Wales at Aberystwyth.

Just as there is regional specialization between universities, so there is concentration in a limited number of departments for the teaching of certain techniques and of some of the more specialized branches of the subject. Most geographers learn at least the elements of cartography and surveying in their undergraduate courses, but advanced work in both these branches is increasingly concentrated in departments that have the staff with the appropriate expertise, and the costly equipment necessary, notably University College, Swansea and Glasgow. Biogeography, concerned with the geographical study of vegetation and soils, was until a few years ago taught effectively in only a handful of departments, but undergraduate courses are now available much more widely, though the scale of laboratory provision varies considerably. Courses in economic geography are given in all departments, though only in a few is there a really close link with national and regional planners, as in the London School of Economics, and at Newcastle and Nottingham. The Univerisity of Reading Department of Geography now has research funds for work in this field of applied geography through its links with the Centre for Environmental Studies.

Reference has already been made to the general teaching in cartography given in most departments. When *Maps and Diagrams* by F. J. Monkhouse and H. R. Wilkinson was first published in 1951, it expounded what was then almost a pioneer course in the university where they were then teaching: and it has subsequently had a very considerable influence upon teaching elsewhere. In rather similar ways there has been a tremendous development in the teaching and use of quantitative methods in both physical and human geography, during the 1960s. The pioneer work by S. Gregory (then at Liverpool, though now at Sheffield), *Statistical Methods and the Geographer* (1961), has been followed by many other books including those by R. J. Chorley (of Cambridge), P. Haggett (Bristol, formerly at Cambridge) and J. P. Cole and Cuchlaine A. M. King (of Nottingham). Equipment for quantitative work in geography, at undergraduate, postgraduate and staff research level, is becoming increasingly available in departments of geography notably perhaps at Newcastle and Bristol, and in London at the School of Economics and University College. Nearly all universities now have their own computer, or easy access to one; even a limited number of undergraduate dissertations now depend on the use of the computer.

The dissertation is an important part of the work of most departments of geography. The necessary field-work is usually done during the last summer vacation before a student's final examinations and the dissertation, or extended essay, or geographical description, as it is variously called, gives the examiner a valuable indication of a candidate's ability to work on his own, with only limited supervision, and to link up field-work with library investigations and discussions with people with local or expert knowledge. Topics vary from the comprehensive account of a limited area insisted upon by some departments to a geographical study of a particular, and sometimes highly specialized, subject. Apart from the value of the discipline of such a study to the individual, a dissertation serves a useful purpose by revealing which students are most likely to make a success of postgraduate research work.

The teaching methods adopted by departments vary according to tradition, the availability of teachers, the staff/student ratio, and the size of departments. Increasingly, lecture courses are supplemented nearly everywhere by tutorials (one, two, three or more students with an individual teacher) or seminars (where larger numbers discuss a specific theme). All departments encourage students to read as widely as possible, emphasizing in particular recent literature, especially that published in geographical and other periodicals. Tutorials and seminars permit the consideration of the meaning and relevance of those papers that have not been specifically discussed in lectures. Group field-work is organized in various ways, ranging from the half-day tour into the area surrounding the department to the week or ten days spent in a selected area farther afield, often during the Easter vacation. During these longer excursions many different activities are undertaken, and a considerable part of the total time will be devoted to individual projects in several different branches of geography. Many departments regularly take at least one group of students abroad in order to familiarize them with an entirely different kind of physical and social environment from that of the British Isles.

Geography has had a long and honoured history, and its progress in British universities has been especially marked in recent years. There are still a few universities, mainly those newly founded and others not yet fully established, where there are no departments of geography, but this lack is more than offset by the size and vigour of many other departments elsewhere. The strength of the subject is also revealed by the growth and activity of its professional societies which include the Royal Geographical Society, the Royal Scottish Geographical Society, the Geographical Association, and the Institute of British Geographers, as well as the Geographical Section of the British Association for the Advancement of Science. Research is fostered in many university departments, and financial support is given by the Natural Environment Research Council and by the Social Science Research Council. Geographers play an increasing role in national affairs, through membership of government committees and of regional economic councils, and by their work as consultants in special fields. For several years the key post of Chief Planner at the Ministry of Housing and Local Government was held by a geographer who is now Professor of Town and Country Planning in the University of Sheffield. There is no lack of outlets at all levels for trained geographers, with especially large groups going into the teaching profession and planning. It is small wonder, therefore, that there is such a demand by geographers for places in universities, in colleges of advanced technology, in colleges of education, and in many other institutions of higher or further education. This situation is likely to continue as long as geography remains the popular subject that it appears to be in many sixth forms, to judge from the large numbers who take the subject at Advanced Level as well as from the enthusiasm of many of those who have studied the subject in some depth for a considerable period.

The U.C.C.A. form-filler needs, therefore, to be both patient and optimistic; and this chapter will, it is hoped, have suggested to him that the opportunities are numerous and varied and that the effort involved is likely to be rewarded. There are so many places in which the subject is attractively and effectively taught today that the choice of a specific department may not matter particularly—certainly not as much as it did in the past. Ordinary common sense and special knowledge should, of course, be used wherever possible, but failure to achieve one's first choice—or even, for that matter, one's sixth choice—need not mean frustration, let alone disaster. The remarkable but very satisfying experience of most university teachers is the discovery that the majority of their students are well content with the department and the university in which they are studying. Most geography students appear to profit from their courses, to enjoy the many sides of the life of their university, and to develop considerable affection and loyalty for their department. This seems to be as true of those whose original entry was by way of the U.C.C.A. clearing-house procedure after the publication of the Advanced Level results as it is of those who go to their first-choice university. There is as yet no problem of employment, even though there are now as many as 1,300 new honours graduates in geography every year. It is a subject with many growth points and a host of enthusiastic teachers, and there will always be room for the keen and determined student of a subject that has throughout its history been noted for its dynamic qualities and for its readiness to adapt to changing circumstances and new conditions.

FURTHER READING

How to apply for Admission to a University, annual, Universities Central Council on Admissions, Cheltenham.

A Compendium of University Entrance Requirements for First Degree Courses in the United Kingdom, annual, Association of Commonwealth Universities for the Committee of Vice-Chancellors and Principals of the Universities of the United Kingdom, London.

Commonwealth Universities Yearbook, annual, Association of Commonwealth Universities, London.

Course Comparison Booklets, annual, Careers Research and Advisory Centre, Cambridge.

JACKSON, C. I., *Degrees in Geography at British Universities*, Royal Geographical Society, London.

12

Careers in Geography

J. Oliver

Geography as a university discipline has displayed great vitality and rapid change in recent years both in its content and scope. The advance towards new frontiers and the modification of old divisions have opened up a wide range of career possibilities to those graduating in geography. This change from pre-war conditions is not yet fully appreciated even by those whose main concern is careers advice.

There are two major sources of information which reveal this wide range. Table I (p. 177) summarizes data collected by the Geographical Association by means of a questionnaire circulated annually to university departments of geography. The strength of this source is that its classification was designed by geographers and is relevant to the main careers that geographers pursue. Unfortunately only three quarters of those graduating in the period 1962-7 are covered by the returns, and not all of those returned were in known occupations. Nevertheless, the information does provide an adequate statistical sample.

The second source of information is contained in the University Grants Committee's annual returns of employment immediately after first graduation. These data have the disadvantage that the classification is designed for all graduates, and therefore many categories of interest to geographers are obscured. Furthermore, the data deal only with the situation immediately after graduation. The Geographical Association's information shows that approximately a half of those graduating proceed to further training, and we really need a return of employment occupation about three years after first graduation in order to obtain a completely accurate picture.

Thus, although an identical six-year period has been selected, detailed comparison between the two sources of information is not easy to make with the different classification systems employed. What the tables do reveal quite clearly, however, is the wide range of career possibilities now open to the geographer. We can also gain some idea of the relative distribution of geography graduates between the different outlets.

If we group together the various possibilities in schools, colleges and universities, the most important major category is undoubtedly teaching; but whereas before the Second World War nearly all geography graduates were absorbed by teaching, only half of today's graduates seek a career in this field. As an educational subject geography has a valuable role to play in a school curriculum. The importance of an intelligent awareness of the world and of the similarities and contrasts between different parts of it need no emphasis. Geography can obviously meet this need. It is a fundamental school subject along with English, History, Mathematics and Science. Teaching is an entirely worthwhile, and to those who approach it in the right manner, a fully satisfying career. Good geographers in the making and the well-informed citizens of the future need good geography teachers.

A not inconsiderable number of graduates still enter the teaching profession directly after obtaining their first degree, but the majority spend a postgraduate year in a university department or institute of education qualifying for the Diploma in Education. Those in colleges of education follow a scheme in which professional and academic training proceed side-by-side either for a Certificate of Education or for the new Bachelor of Education degree.

Teaching geography at a more advanced level in colleges of education, technical colleges and universities requires not only a good honours degree but also, in the majority of cases, a special interest in a particular branch of the subject. The University Grants Committee statistics on postgraduate employment show that, for the U.K. in the six-year period to which they refer, 53 per cent of those with higher degrees in geography found posts in universities or similar institutions. An analysis (Table IV p. 180) of university lecturing posts advertised in the British press over a period of twelve years from 1957-68 gives an idea of the relative strength of the demand in various fields.

A small category of opportunities of a rather different nature, but essentially teaching in character, includes the posts of field assistants or wardens, or similar openings at the centres of the Field Studies Council (see appendix). Many of these require a geographical background, although those also able to offer additional qualifications in other earth and life sciences, or environmental studies, are likely to be best placed in the competition.

The second major category of employment is local government, where many geography graduates are attracted to town and country planning. The qualifications for a chartered town planner are reminiscent of a statement on the geographer's field of study. 'For work in Development Plans he must acquire an intimate knowledge of the physical features, natural resources and the economic and social character of the place in which he is working. He must be equally conversant with surveys of existing land-use and able to utilize different types of traffic survey and present his data effectively and lucidly' (*Town Planning as a Career*). Like the geographer, the town and country planner has to appreciate the interplay of circumstances that result in the character, distribution and organization of social and economic systems, and then has to make beneficial and acceptable proposals for the maintenance, modification or improvements of such systems. Inevitably, the understanding entails the study of human society and its functioning, but it demands in addition a consideration of the spatial framework of natural conditions within which the society operates. The geographer's training is clearly very appropriate, and although he may not be conversant with all that is relevant in civil engineering, architecture, or legal matters, these are gaps that can be filled.

Further professional qualifications are almost a *sine qua non* for advancement. Associate membership of the Town

Planning Institute is the qualification usually sought. It can be achieved in two main ways. On the one hand in-service training in a county or municipal planning department provides opportunities for practical experience. In the better authorities favourable study conditions, perhaps in conjunction with block or day-release to a local technical college, enable the trainee-planner to prepare for the professional examinations. The other line of training is through post-graduate degree or diploma courses in planning which are available in various universities or technical colleges; these may give exemption from the Town Planning Institute's final examinations. Associate membership of the Institute also requires that applicants should have at least two years' practical experience in town planning.

Geography graduates with interests in fields closely related to the problems likely to concern the planner are in the best position to be considered for planning posts, but those geographers without specialist labels are by no means excluded. If a degree course has included more concentrated and advanced work in urban and settlement geography, demography, quantitative economic geography, locational analysis or regional science, then career prospects are enhanced; but many planning authorities are alive to the fact that any geographer can offer valuable attributes through the breadth of his studies and his mental awareness of the inter-relationships between the many component parts of an investigation within a spatial framework.

It is sometimes felt that those who have pursued a science-based course with interests in such fields as geomorphology, climatology, hydrology and pedology will not have much prospect of planning appointments. This is not so for this type of expertise is in fact needed in planning work—though perhaps less frequently than those fields with a social science bias. Those qualified in land-use surveys, not, of course, limited to rural areas, will have a training which will be particularly attractive for some of the posts offered by planning authorities.

The third major category of employment is the Civil Service where again the career opportunities are very varied. First, there are the openings in non-specialist fields where geographers compete on equal terms with other graduates. The Administrative class, the Diplomatic Service and the Overseas Service (grades 8 and 9) are the main groups. Selection Method I involves examinations in which geography can be offered as an optional subject. Selection Method II depends more on personal qualities, though there is a greater emphasis on academic achievement than is usually displayed in appointments in the private sector. Further openings for geography graduates occur in the A class and departmental specialists of the Government Communication Headquarters (usually with a mathematical background and successful research record), Tax Inspectors, the Cadet Grade of the Ministry of Employment and Productivity, Assistant Postal Controllers of the Post Office, Officers of Customs and Excise, the Ministry of Employment and the Executive Class of the Home Civil Service. A geography graduate would also be acceptable for posts in the Factory Inspectorate in the Ministry of Employment, where his position would be strengthened if his background included a bias towards the physical sciences or mathematics.

The geographer can also compete for the Linked Research Classes of the Home Civil Service. Here are included posts in the Ministry of Defence Joint Intelligence Bureau, the Home Office, the Board of Trade, Ministry of Agriculture, Fisheries and Food, Ministry of Housing and Local Government including the Welsh Office and the Scottish Development Department, and the Ministries of Health and Employment. The geographer may also be considered, if he has additional qualifications in economics and/or a foreign language, for some of the Research Assistant posts in the Diplomatic Service.

On occasions a geography graduate can offer qualifications which suit him for work in Government research organizations such as the Hydraulics Research Station, the Institute of Hydrology or the Road Research Laboratory. There may be possibilities for the recruitment of geography graduates with special qualifications by the Ministry of Overseas Development on behalf of independent Commonwealth Governments and, more rarely, on behalf of foreign Governments. In the Land Resources Division of the Directorate of Overseas Surveys, integrated land resource appraisals offer opportunities for contributions from geomorphologists, pedologists, climatologists, hydrologists, biogeographers and agricultural geographers and agricultural economists.

We are now, however, entering a fourth major category where the career opportunities largely depend on the specialist qualifications that the geographer may have acquired through his study of one or more of the systematic branches of the subject. The number of vacancies which occur each year in such specialist fields are limited, and the individual keen on following a particular line may find it advisable or necessary to acquire further specialist qualifications whilst waiting for an opening.

The geographer's expertise in the use of maps might be expected to lead into careers associated with cartography, topographical surveying or photogrammetry. It must be appreciated, however, that such work requires a mathematical competence, at least up to the Advanced Level of the General Certificate of Education, together with a technical knowledge. In the majority of cases the content of a normal honours degree in geography is insufficient in itself to qualify the graduate for such work. Generally elegibility for such careers depends on having pursued a special option course in cartography, surveying or photogrammetry within an honours degree, such as the special cartography and survey courses offered at Glasgow, Newcastle and Swansea, or the special survey courses at Oxford and Cambridge, or the post-graduate Diploma in Land Surveying at University College London.

The Directorate of Military Surveys and the Directorate of Overseas Surveys offer certain openings in survey work. The Directorate of Overseas Surveys, which is part of the Home Civil Service and has its headquarters at Tolworth, is concerned with geodetic and topographical surveying in Commonwealth countries. For acceptance into topographical surveying a further year of training at the Military School of Survey, Newbury, is required. Although the number varies, about six male graduates are absorbed each year. The potential demand for those with a training in topographical survey is considerable. Private professional surveying firms engaged in work on roads, bridges, tunnels or harbours offer opportunities for those with a suitable survey training, and geography graduates have entered such fields.

In the case of cartography, a clear distinction should be made between draughtsmanship and cartography. Competence in the latter is the primary aim of a cartographical course, though gifted individuals may develop skills in draughting alongside their main expertise in map compilation. Within the framework of honours schemes in geography, it is possible to pursue special option courses in cartography at Edinburgh, Hull, Keele, Leeds, Leicester, London, Manchester and Swansea, whilst diploma courses are offered at Glasgow and Swansea and an M.Sc. course at Newcastle. Portsmouth College of Technology also offers a special course in cartography as part of the external London degree scheme.

Except at the non-graduate level, entry into the Ordnance Survey has been through the Royal Engineers, so that specialist cartographical training acquired in a geographical course has not in the past assisted entry. It has now been decided, however, that 31 out of the 42 senior posts will be open to civilians. Graduates, after a period of some years' service in the Directorate of Overseas Surveys, could transfer to the Ordnance Survey. For the remaining posts the geography graduate could seek a commission in the Royal Engineers direct from university.

Cartographical openings also occur in the Directorates of Military and Overseas Surveys. Graduates with an honours degree in geography, or a degree in other subjects with geography as part of the scheme, can find appointments as Assistant Map Officers in the Directorate of Overseas Surveys. These officers work on the base material for map compilation, including survey and photogrammetric data, and some combine computer work with their map work. In the Directorate of Military Surveys, graduates applying for posts as Map Research Officers will be in the best position if they have included cartography or surveying in their courses, and have a knowledge of mathematics and of at least one foreign language. Recently the supply of suitably qualified applicants has not met the demand. Other opportunities for geography graduates occur in the Hydrographic Department of the Ministry of Defence—here Civil Hydrographic Officers are concerned with the preparation and correction of Admiralty charts.

From time to time careers involving cartography are advertised by commercial firms, especially map and atlas publishers such as George Philip and the Clarendon Press, or other bodies such as the Map Room of the British Museum or the Anti-Locust Research Centre. It may often be the case with private employers that such opportunities are negotiated on an individual basis and do not, therefore, appear in public advertisements. A further outlet for those with cartographic and draughting experience exists in university departments of geography, some of which have graduate cartographers and many a small cartographic team.

The increasing sophistication of photogrammetric map-making techniques, limitations on the instructional time available, and the capital cost of much of the equipment used, restrict the opportunity to teach most geographers photogrammetry to the level that may be desired by potential employers. Geography graduates who have already pursued some form of special option course involving photogrammetry in their first degree, and who have sufficient mathematical knowledge, are well placed to continue with a post-graduate course in photogrammetry, and can then compete for posts available in aerial survey at such firms as Hunting Surveys and Consultants Ltd and Fairey Surveys Ltd.

In a small number of geography departments optional specialized courses in pedology form part of the first-degree scheme. These provide a more advanced training which enables some graduates with good honours degrees to compete for posts of Scientific Officer in the Soil Survey of England and Wales and the Soil Survey of Scotland. There may also be openings in the Experimental Officer grade for those with lower levels of degree attainment. Where the task is one of field mapping and the consideration of physical processes in soil development, the geographer with the right background is often acceptable. His prospects for successful entry to the posts mentioned would be enhanced by a science-based training which included one or more of the following subjects: Geology, Botany, Ecology, Chemistry. Where resource surveys include work on soil mapping and classification suitably equipped geography graduates may find openings with such organizations as Hunting Surveys and Consultants Ltd.

Honours geography graduates are eligible for the occasional vacancies of Research Officer that arise in the Agricultural Land Service of the Ministry of Agriculture, Fisheries and Food. These posts occur at regional centres which are often located at universities with Schools of Agriculture. Although the National Agricultural Advisiory Service finds most of its recruits from agriculturalists and biologists, B.Sc. graduates in geography with the appropriate subsidiary subjects may find occasional openings.

The Nature Conservancy from time to time provides opportunities for geographers who have, at first-degree level or in subsequent post-graduate research, acquired the appropriate scientific background. For work in the Nature Reserves maintained by the Conservancy, those with a biological training are usually required, but occasionally a geography graduate who has developed interests in the direction of geomorphology, climatology, hydrology, pedology, biogeography, or land-use studies will have the right qualifications. Those geographers who have pursued a conservation course such as that at University College London will be well placed for this type of work.

Occasionally posts arise with the Countryside Commision as Assistant Field Advisors, Planning or Research Officers, or with the Planning Committees of the several national parks as field staff, or with the National Trust or the Council for the Protection of Rural England. On rare occasions other opportunities may occur, such as the post advertised for a historical geographer to undertake investigation into the land-use history of grasslands and heathlands at Abbot's Ripton.

Some departments of geography have been developing specialized courses in hydrology, and have been building up links with River Authorities, the Water Resources Board at Reading, and the Institute of Hydrology (formerly the Hydrological Research Unit) at Wallingford. There are career prospects in this field for those geography graduates who have pursued an optional course in hydrology and have some competence in mathematics. If the individual geographers who enter such posts give satisfaction and demonstrate the usefulness of their training and approach, there are good prospects for further appointments by satisfied authorities. At present there is a shortage of suitably trained hydrologists for the many posts that have to be filled. These posts have increased in number as a result of the 1963 Water Resources Act and the pressing need to resolve our problems of water supply, flood protection and water conservation. The keen hydrologist/geographer would be well advised to supplement his degree in geography with a post-graduate course in hydrology. Such courses are available at Imperial College, London, Newcastle and Birmingham.

Mention should also be made of meteorology and climatology. Having had an interest in these disciplines stimulated by specialist courses in their geographical studies, some students aspire to a career in them. Those geographers who have appropriate specialist interests and subsidiary qualifications in mathematics and physics may be successful in the grade of Experimental Officer in the Meteorological Office, but, for the higher grade of Scientific Officer, specialist knowledge of mathematics and physics is a necessity. Women geography graduates (preferably with honours) with similar subsidiary qualifications can pursue a meteorological career in the Instructor Branch of the Women's Royal Naval Service, after the successful completion of a training course at the R.N. Air Station at Culrose, Cornwall.

Our fifth major category covers opportunities in the economic, commercial, business and industrial worlds. Once again the specialist interest acquired through the study of one or more of the systematic branches of geography can be of considerable assistance. Appropriate joint degrees can also be advantageous. Economic geographers, more particularly those who have adopted a quantitative approach in their studies, can compete with economists for posts concerned with economic surveys or agricultural economics. Their employment potential in this respect is strengthened if economics has been associated with the geography course either as a subsidiary or as a joint degree subject. Both the disciplines are complementary in this type of work. Indeed, once again the usefulness of the geographer having made studies in other disciplines at university level, or even at the Advanced Level of the G.C.E., is demonstrated. A postgraduate course in business studies is a valuable additional qualification for a geographer wishing to follow a career in economic matters.

A certain amount inevitably depends on the initiative of the individual. For instance, it might be thought unlikely for a geographer to obtain an appointment as an Agricultural Economist with the Civil Service. A geography graduate with subsidiary economics, or with combined honours in economics and geography, might, however, obtain an award from the Ministry of Agriculture, the Agricultural Research Council, or the Milk Marketing Board to pursue a postgraduate course in Agricultural Economics. This in turn could open up the possibilities of a career in the Ministry of Agriculture's Provincial Agricultural Economics Service, in the Marketing Division of the Milk Marketing Board, in a college or university agricultural department or a farm institute, with a farmers' organization, an agricultural merchant, or manufacturer or distributor of agricultural equipment and requirements.

There is keen competition for Cadetships in the Government Economic Service, but a suitable geography graduate is eligible for consideration. Openings in regional planning in the Ministry of Housing may also occur, and the Board of Trade, with its concern in regional planning and industrial location, provides further possibilities. Geography graduates have been employed in research posts by the Economist Intelligence Unit. A few professional geographers have successfully established themselves as economic planning consultants. In other directions geography graduates can find opportunities in the fields of transport management, shipping information services, freight despatch and export departments in individual firms. Such posts have to be sought out and may not be advertised in association with a request for a geography degree. British Rail, however, recognizes geography as one of a number of desirable subject qualifications for posts involved with traffic surveys and forecasts, commercial appraisals and market research.

Resource assessment is becoming increasingly important in our world today. The growing need for conservation and the search for new resources of raw materials is essentially geographical in nature. Evaluation of resources is a multivariate problem, and geographers have valuable contributions to make in dealing with this problem. So far few national or international bodies, including the various agencies of the United Nations, who are involved with resource assessment and management, have sought geography graduates as such, but the geographer who has the appropriate specialist interest or post-graduate qualification is acceptable.

In the commercial field firms have not, until fairly recently, named specific degrees when recruiting graduates. Geographers must therefore compete in most cases with graduates from other disciplines for many types of non-specialist employment. It is justifiable, however, to claim that, compared with many other graduates, the geographer has a better potential in the world of commerce with his environmental awareness and skill in literacy, numeracy and graphicacy. It is of interest to note, therefore, that the *Cornmarket Directory of Opportunities for Graduates* and the C.R.A.C. publication *Beyond a Degree* have begun to list firms where a degree in geography is particularly welcome. The most recent issues include: Alcan Industries Ltd, Bank of America, Bank of London and South America Ltd, Barclays Bank, British Airport Authority, British Oxygen, British European Airways, British Motor Corporation, British Rail, British Transport Docks Board, C. and A. Modes, Cadburys, Dorman Long Ltd, Electricity Supply Industry, Economist Intelligence Bureau, Hall-Thermotank Ltd, Heinz, I.B.M., Marks and Spencers, McDougall Ltd, Metal Box, Perkins Manufacturing, Parkinson Cowan, Rank-Hovis, Richard Thomas and Baldwins, Tube Investments.

On the assumption that their interests in different parts

of the world would have been stimulated by the very nature of their degree studies, it might be anticipated that geography graduates would find many openings overseas. This belief is strengthened by the fact that geographers frequently appear as the organizers or members of overseas expeditions of exploration. A number of suitably trained geographers have found work abroad, especially in Africa and parts of Southern Asia, as topographical surveyors. Other geographers, with a good scientific background and strong personal qualities, have joined the British Antarctic Survey in the capacity of surveyors, geomorphologists, glaciologists, or meteorological observers. These are short-term contracts, however, and can only be looked on as providing valuable practical experience before proceeding to a final career. Less often geographers have found employment in Africa in regional surveys and land planning. The British Council has also appointed geography graduates to its various posts from time to time, but additional language qualifications are clearly of importance here. In Australia, the Commonwealth Scientific and Industrial Research Organization has a geographical division in the Department of Land Use and Regional Survey in which geomorphologists and land-use experts are employed. In the U.S.A., geographers can find a much wider range of employment than in Britain. Official openings occur in the National Park Service of the Department of the Interior in addition to other executive agencies of the Federal Government. In Canada, geographers as such are employed in the government service in the assembly, collation and processing of geographical information useful to the economic, commercial and social life of the country.

As with the home situation, however, overseas employment has been mainly in teaching, especially in the universities. Many openings in North America, Africa, Australia and New Zealand and to a limited extent in Southern Asia have arisen in the educational explosion of the last two decades. One point should be appreciated, and that is that, once overseas, the teaching graduate should be prepared for a long period if not for a permanent life abroad. Return to work in Britain is not always easy to arrange. Voluntary service overseas, however, will give graduates the opportunity to judge whether they wish to find a permanent appointment overseas.

Perhaps one of the most important pieces of advice to offer to the geographer in search of a career is the need to show initiative and enterprise and a willingness to acquire further skills or training beyond his systematic studies. The Geographical Association enquiry and the University Grants Committee returns both reveal an incredibly wide range of possibilities. In addition to the activities that we have specifically mentioned in the preceding discussion, geographers have occasionally entered the armed services, the police services and the Church. They have taken up social work both at home and overseas, trained as librarians and museum curators, entered the world of publishing, radio and television, and taken up legal and accountancy work. The records even reveal one professional footballer amongst a number who have combined physical education with their geography, whilst yet others were employed by the National Union of Students and a brewery! The range is indeed wide, but there are many fields yet to conquer. As Professor M. Cole commended: 'It remains for geographers with energy and foresight to seek a footing in the City in banking and insurance, the stock market and in finance houses and there to bring their geographical skill to bear on the new disciplines they must learn.'

It is for the student himself, guided by his specialist interests, to make the choice. With a good geography degree and initiative he should have no difficulty in finding suitable employment. In the words of the Rt. Hon. Harold Wilson, writing of the career prospects of graduates.

> 'It is of the greatest importance that this growing body of students should, on completing their first-degree studies, find employment opportunities in which they can make the most of their abilities and of their education. They are needed in industry and commerce, in national and local administration, in teaching and in research and in other essential departments of our national life.'

Table I

Distribution of first-degree geography graduates in main employment groups during the period 1962–7. Based on the annual questionnaire distributed by the Geographical Association to university departments of geography and covering 4,250 cases (about three quarters of those graduating with honours in geography during the period).

Type of occupation entered into immediately after graduation	*Percentage of total*
School teaching (without a post-graduate training)	10.8
Teacher training (other than primary school training)	32.7
University and further education teaching	1.8
Research or further academic study	13.4
Town and country planning	6.6
Research officers (housing and local government)	0.7
Civil Service, Colonial Service and local government	3.0
Cartography, surveying (other than for local government or Civil Service)	0.8
Commerce (insurance, shipping, banking, etc.)	2.9
Management and industry	2.8
Estate management, agriculture, building, civil engineering, contractors, mining, distributive trades, nationalized industries, public utilities, market and industrial research	2.3
Armed services, police	0.6
Training for social work and the Church, social and personnel work	1.6
Librarianship training	0.4
Secretarial training and secretarial work	1.7
Publishing, radio and television	0.5
Miscellaneous	7.6
Not Known	9.8
Classified as using geography professionally	71.2
Classified as using geography as a Liberal Study	19.0
Classified as proceeding to further training	53.6

Table II

Employment distribution of first-degree graduates in geography immediately after graduation over the period 1962–7 according to U.G.C. returns. These figures cover about two fifths of those graduating—over a half continue with further training. Percentage figures refer to total employed in the United Kingdom.

	Men	%	*Women*	%	*Total*	%
Schools	161	14.5	213	35.7	374	22.0
Technical colleges and further education	15	1.4	3	0.5	18	1.1
Universities	17	1.5	23	3.8	40	2.3
Local government and hospital services	312	28.0	145	24.0	457	26.0
Home Civil Service, Diplomatic Service	73	6.6	70	11.7	143	8.4
Agriculture, forestry and horticulture	2	—	0	—	2	—
National Coal Board, other mining & quarrying	3	—	2	—	5	—
Oil industry	17	1.5	5	0.8	22	1.3
Chemical and allied industries	31	2.8	15	2.5	46	2.7
Engineering and allied industries	139	12.5	16	2.6	155	9.1
Other manufacturing industries	81	7.3	8	1.3	89	5.2
Builders, contractors, civil engineers, architects	17	1.5	5	0.8	22	1.3
Public utility, atomic energy, transport undertakings	49	4.4	5	0.8	54	3.2
Accountancy and commerce	125	11.3	44	7.3	169	9.9
Armed services	17	1.5	5	0.8	22	1.3
Legal profession	5	—	1	—	6	—
Book, magazine, newspaper publishing	15	1.3	16	2.6	31	1.8
Cultural organizations	5	—	4	—	9	—
Other employment, not specified	20	1.8	16	2.6	36	2.1
Total employed in U.K.	1104	—	596	—	1700	
Employed abroad, seeking employment & details unknown	304		170		474	
Total covered by survey	1408		766		2174	

Table III

An alternative breakdown of the statistics in Table II showing types of work undertaken by first-degree geography graduates entering employment immediately after graduation during the period 1962–7. Extracted from U.G.C. returns; see also caption to Table II.

Type of Work	*Men*	*Women*	*Total*	%
Teaching	193	226	419	24.6
Scientific research	18	8	26	1.5
Design and development	173	69	242	14.2
General traineeship	226	27	253	14.9
Postgraduate apprenticeships	81	27	108	6.4
Scientific analysis and investigation	20	17	37	2.2
Technical and advisory work	7	8	15	0.9
General administration	68	39	107	6.3
Production operation and maintenance	23	3	26	1.5
Buying and selling	73	13	86	5.1
Statistics and economics	86	76	162	9.5
Accountancy and finance	82	2	84	4.9
Legal and patent work	5	1	6	0.4
Editorial, information and journalism	14	25	39	2.3
Museums, library and archive work	7	19	26	1.5
Film, radio, television and stage	1	0	1	—
Personnel, safety and establishments work	13	11	24	1.4
Social work	4	8	12	0.7
Secretarial and clerical	2	15	17	1.0
Others	8	2	10	0.6
Totals	1104	596	1700	

Table IV

An analysis of the fields of interest of university lecturing posts in geography advertised in *The Times* during the period 1957–68.

Description of Post	*Percentage of total number of university lecturing posts advertised*		
Special interests desired	*Home universities*	*Overseas universities*	*Combined*
1. Social, economic, political, urban, agricultural geography, land-use	13.3	9.9	23.2
2. Historical geography	1.4	0.8	2.3
3. Regional geography, regional planning	6.2	2.3	8.5
4. Quantitative geography	1.7	1.6	3.4
5. Cartography, surveying, photogrammetry, mathematical geography	4.6	2.1	6.7
6. Biogeography, climatology, geomorphology, hydrology, pedology, physical geography in general	12.3	11.9	24.2
7. Miscellaneous, specified qualifications not included above	2.6	1.4	4.0
8. No special interests other than a qualification in geography required	15.3	12.4	27.7
	57.4	42.4	100.

FURTHER READING

Careers for Geographers, Royal Geographical Society, London.

Cornmarket Directory of Opportunities for Graduates, annual, Cornmarket Press, London.

Civil Service Posts for Graduates, annual, Civil Service Commission, London.

The Scientific Civil Service, annual, Civil Service Commission, London.

Town Planning as a Career, The Town Planning Institute, London.

Beyond a Degree, A Yearbook of Education and Training opportunities. Careers Research and Advisory Centre, Cambridge.

United Kingdom Postgraduate Awards, annual, Association of Commonwealth Universities, London.

COLE, MONICA, *Careers for Geographers*, Geography Vol. 47, 1962.

OLIVER, J. *Formula for Success*, Geographical Magazine, February 1970.

Appendix: Additional Information

W. G. V. Balchin

The Universities Central Council on Admissions

The Universities Central Council on Admissions (U.C.C.A.) covers nearly all the universities of the United Kingdom and was formed to facilitate the entry of intending students to universities. The scheme is concerned only with full-time courses leading to a first degree or first diploma at universities within the United Kingdom and applies to all candidates whether their permanent home address is in the United Kingdom or overseas.

Candidates for admission to a university are required to complete a form obtainable from heads of schools or principals of colleges of education or, where the applicant is not in full time education, from U.C.C.A. This must be submitted during the period 1 September to 15 December for entry in the following October, except for Oxford and Cambridge where special arrangements exist. Candidates in 1970 and thereafter are allowed to list up to five universities in order of preference, and their application form will be duplicated and forwarded to all five institutions. The selected universities then consider the applications and through U.C.C.A. either make conditional or unconditional offers to the candidates as appropriate, or they will put the application on a waiting list, or they may reject the application. Most offers are conditional upon certain grades being reached in the Advanced Level examinations, and normally candidates will not know for certain that they have a university place until late in August when the Advanced Level examination results are published. Candidates who have failed to secure any conditional offer or who fail to reach the required standard of conditional offers which have been made can be reconsidered in the 'clearing-up procedure' in September when all universities inform U.C.C.A. of their vacant places. Full details of the U.C.C.A. arrangements may be obtained on application to the Universities Central Council on Admissions, P.O. Box 28, Cheltenham, Gloucestershire.

Books

Over fifty firms in the United Kingdom are now engaged in the publication of geographical books, texts and journals. It is impossible within the confines of this introductory volume to cite even a select list concerned with the content of the subject, but the intending student may like to know of the geographical lists published by leading booksellers such as: Blackwell's, Broad Street, Oxford; Dillon's, Malet Street, London, W.C.1; Heffer's, Petty Cury, Cambridge.

Maps

As geographers are particularly concerned with the maps, the main official publishers and reference collections for Great Britain and Ireland are noted below:

The Ordnance Survey at Romsey Road, Maybush, Southampton is responsible for the basic surveys and topographic maps of Great Britain. It also publishes a great variety of other maps for various government departments and agencies. Ordnance Survey publications can be obtained from any Ordnance Survey agent and from most booksellers. A list of agents and a catalogue will be sent on request.

The Institute of Geological Sciences, Exhibition Road, South Kensington, London, S.W.7 is responsible for the geological maps of Great Britain. The maps are listed in a catalogue available on request from the Director of the Institute of Geological Sciences or from the Ordnance Survey which publishes the maps. Geological Survey maps are obtainable in the same way as other Ordnance Survey publications.

The Soil Survey of England and Wales, at Rothamsted Experimental Station, Harpenden, Herts., and *The Soil Survey of Scotland,* at the Macaulay Institute for Soil Research, Craigiebuckler, Aberdeen are responsible for the soil survey maps of Great Britain. A list of maps currently available is obtainable on request from the publishers, the Ordnance Survey.

Land Use Survey Maps of Britain are prepared by the Second Land Utilisation Survey of Britain at King's College Strand, London, W.C.2. A catalogue of publications may be obtained on request. Maps may be purchased either from the Second Land Utilisation Survey at King's College, London or from Edward Stanford, Ltd, 12-14 Long Acre, London W.C.2.

Ireland. Enquiries about maps for Northern Ireland should be addressed to The Chief Survey Officer, Ministry of Finance, Ordnance Survey, Ladas Drive, Belfast. Enquires about maps of the Irish Republic should be addressed to The Assistant Director of Survey, The Ordnance Survey, Phoenix Park, Dublin.

Reference Collections. All university departments of geography maintain selective reference collections of maps. Comprehensive reference collections will be found at the Royal Geographical Society, Kensington Gore, London, S.W.7, and at the copyright libraries, more especially the British Museum, Bloomsbury, London, W.C.1, the Bodleian Library, Oxford and the University Library, Cambridge.

Field Studies

There are a number of bodies concerned with field studies who provide assistance to geographers in various ways:

The Field Studies Council was founded in 1943 as the Council for the Promotion of Field Studies by Mr F. H. C. Butler to encourage every branch of field work. It achieves its main aim through a number of field centres which provide accommodation for field groups and also offer field courses. Details of the annual programmes at the centres may be obtained on application to the headquarters of the Field Studies Council at 9 Devereux Court, Strand, London, W.C.2 or to the Warden in charge at each centre. The field centres are located at: Dale Fort, Haverfordwest, Pembrokeshire; Flatford Mill, East Bergholt, Colchester, Essex; Juniper Hall, Dorking, Surrey; Malham Tarn, Settle, Yorkshire; Orielton, Pembroke, Pembrokeshire; Preston Montford, Shrewsbury, Shropshire; Slapton Ley, Slapton, Kingsbridge, Devon; Rhyd-y-Creuau, Betws-y-coed, Caernarvonshire (Drapers Field Centre); Nettlecombe Court, Williton, Taunton, Somerset. (Leonard Wills Field Centre).

The Scottish Field Studies Association offers similar facilities in Scotland mainly through its Kindrogan Field Centre at Enochdhu, Blairgowrie, Perthshire. Other courses are arranged in co-operation with the Holiday Fellowship. The headquarters of the Scottish Field Studies Association are at 141 Bath Street, Glasgow C.2.

The Geographical Field Group is a non-profit-making organization which arranges geographical field studies both in Britain and on the continent of Europe. Its headquarters are c/o the Department of Geography, The University, Nottingham, although it has no formal connection with the university. Enquiries should be addressed to the Hon. Secretary.

The Geographical Association also organizes fieldwork and study courses. The activities of the Association are more fully discussed below in the section on Societies.

The Youth Hostels Association (*England and Wales*) (Trevelyan House, St. Albans, Herts.) and *The Scottish Y.H.A.* (7 Gleve Crescent, Stirling) provide field study facilities at some of their hostels.

The Nature Conservancy (19 Belgrave Square, London, S.W.1), now part of the Natural Environment Research

Council, is the government body responsible for nature conservation and related research and for providing scientific advice on these subjects.

The Council for Nature (Zoological Gardens, Regent's Park, London, N.W.1) is the representative body of the voluntary natural history and conservation movement in Britain. The Council can provide the addresses of local and national natural history, conservation and countryside organizations, and answers enquiries on all aspects of British wild life.

The County Naturalists and Conservation Trusts (The Manor House, Alford, Lincs.) are the principal local voluntary organizations concerned with nature conservation and the provision of nature reserves. There are now 36 trusts in Britain, including the *Scottish Wildlife Trust*, linked together by the Society for the Promotion of Nature Reserves. They manage more than 350 nature reserves, some specially for educational use. They can often advise on the other sites suitable for field studies and welcome information about them.

Learned Societies

The Royal Geographical Society is one of the oldest geographical societies in the world, and is the chief British centre of exploration and geographical information. Its object, as defined in its Royal Charter granted by Queen Victoria, is the advancement of geographical science. In common with many other early learned societies, the R.G.S., as it has come to be known, can trace its descent from a dining club. The Raleigh Dining Club was founded in February 1827; each division of the world was represented by at least one member with first-hand knowledge. From this dining club the Royal Geographical Society emerged under the patronage of King William IV in 1830. In the following year it absorbed an even earlier society, the African Association, which had been founded by Sir Joseph Banks in 1788 for the promotion of travel and discovery in Africa.

To read the history of the Society is to read the history

of British geographical exploration and discovery in the nineteenth and twentieth centuries—Livingstone, Stanley, Scott, Younghusband, Shackleton, Hunt, Hillary, Fuchs, and many more well-known names have contributed to its activities. The Society has also helped to launch and develop geography as an educational and academic subject in Great Britain through the work of Mackinder, Scott-Keltie, E. G. R. Taylor, Stamp and many other Fellows.

The Society welcomes to its Fellowship all those who are anxious to promote its objects, even though they may not themselves be able to take an active part in exploration or may not have published original work. Its membership currently numbers 6,700. Candidates for Fellowship must be proposed by a Fellow of the Society from personal knowledge. The Society also has a class of Associate Membership at a reduced annual fee expressly for those aged 17-20. These members, provided that they have been Associates for three years, are eligible for Fellowship at age 21 without payment of the entrance fee.

Fellows and Associates are entitled to attend all meetings of the Society at which they may introduce guests, and they also have the use of the Society's Library and Map Room. Both of these are unique collections of geographical material. The Library contains well over 100,000 books and the map collection well over 500,000 maps. Fellows or Associates resident in Great Britain may also borrow books by post or in person, but maps must normally be consulted at the Society's house. In addition, all Fellows receive the quarterly *Geographical Journal,* and Associates may choose between the *Journal* and the monthly *Geographical Magazine.* The latter is not published by the Society but was founded with its agreement and support.

The Society's house also contains a museum, a periodicals room, rooms for reading or discussion, drawing offices, a large lecture hall, exhibition space, and expeditionary offices. The address of the Society is 1 Kensington Gore, London, S.W.7 and enquiries should be addressed to the Director and Secretary.

The Royal Scottish Geographical Society was founded in

1884. The decision to form a Scottish geographical society was made at a meeting on 20 July of that year between John George Bartholomew (of the cartographic firm) and Mrs Agnes Livingstone Bruce (daughter of David Livingstone). A committee was formed and the Society launched at a public meeting in Edinburgh on 28 October 1884.

The Society publishes the *Scottish Geographical Magazine* and arranges lecture programmes in Edinburgh and at branches in Aberdeen, Dundee, Dunfermline and Glasgow. It has a cartographic section for the discussion of aspects of modern cartography, and it organizes field work in and tours to other countries. The Society maintains a book library, periodicals library and map library at its Edinburgh headquarters. All books and periodicals and some maps are available for loan.

The Society welcomes to its membership all those who are interested in the study of geography. It has various categories of fellowship and membership, including a reduced fee scheme of associate membership for students at Scottish educational institutions. The headquarters of the Royal Scottish Geographical Society are at 10 Randolph Crescent, Edinburgh 3, and enquiries should be addressed to the Secretary.

The Geographical Association was founded in 1893 largely as a result of the activities of B. B. Dickinson of Rugby School with the aim of furthering the study and teaching of geography. It caters for all categories of educational institutions, from primary and preparatory schools to universities, both in the United Kingdom and abroad. Its services include: the publication of the Association's quarterly journal *Geography* and a large range of special publications, the maintenance of a geographical library, the organization of conferences and courses, and the promotion of local branch activities throughout the country. Its membership now approaches 10,000.

The annual conference takes place in London about the New Year and is generally attended by 700 to 800 members. The programme includes illustrated geographical lectures, discussions, demonstration lessons and symposia on geographical matters. The spring conference, usually attended by 200 to 300 members, is a residential meeting, held annually at a provincial centre about Easter. The programme generally includes excursions and lectures on the geography of the region in which the conference is being held. Residential summer schools and courses in geographical field studies are held both in the United Kingdom and overseas during the summer vacation. In recent years visits have been organized to places as far afield as Canada, the West Indies, and the Middle East.

Local branches exist at a large number of centres throughout Great Britain, and each branch is free to arrange meetings, discussions, lectures, excursions, field work, etc. Full members of the Association are entitled to join any branch provided that they pay the local fee. Other persons who are not full members of the Association may become associates of a branch on payment of the appropriate subscription.

The Association welcomes to its membership all those who are anxious to promote the study and teaching of geography. Full members are entitled to receive *Geography* four times each year, to attend the annual and spring conferences and to participate in summer schools, and they may borrow books either by post or in person from the Association's library of 10,000 volumes. They may also join any local branch of the Association and participate in its activities subject to the payment of the local membership fee. The Association also has a special category of student membership with a reduced subscription. Students in universities and colleges and senior pupils of secondary schools may be enrolled at a privilege rate with the option of *either* receiving *Geography or* using the library. Only *bona fide* full-time students may benefit from this concession, and applications must be accompanied by a certificate of studentship from a headmaster, principal or professor. Associate members of local branches will normally take part only in local activities organized by the branch.

The Association's headquarters are in Sheffield, and in addition to the library and publishing section the building contains extensive offices, a map room and conference rooms. Details of membership, local branches throughout Great Britain and all publications of the Association may be obtained on application to the Administrative Secretary, the Geographical Association, 343 Fulwood Road, Sheffield.

The Institute of British Geographers had its origins in a luncheon party in January 1931, when Dr R. O. Buchanan, Dr H. A. Matthews and Dr S. W. Wooldridge discussed the lack of opportunities for publishing geographical research work. As a result of this discussion a meeting of London geography staff was convened, and from this a report was sent to the heads of university geography departments conference for consideration. After further general meetings in 1932 the I.B.G. was founded with the prime aim of publishing geographical material on the basis of subscription fees. Its membership, therefore, has always been of graduate standing and has largely consisted of dedicated professional and academic geographers plus a substantial number of postgraduate research workers. The growth of interest in geography and of the desire to further its advancement through the publication of research work can be judged from the increase in membership to a present total of over 1,200.

Today the Institute regularly publishes its *Transactions* which contain major research papers, and from time to time it issues monographs and special books. It has recently begun to publish a new quarterly, *Area,* to provide space for short articles on subjects of scholarly interest to geographers. *Area* also serves as a medium for the expression of professional opinions on matters of public interest, and as a channel for the dissemination of news of the activities of members of the I.B.G.

A second aim of the Institute is to further the advancement of the subject by personal contact and discussion between its members. This is achieved through its annual residential conference which migrates among university departments attracting attendances of up to 500, and also through its systematic study groups which meet in locations appropriate to the topics under consideration. Currently there are study groups for agricultural geography, agrarian landscapes, population geography, quantitative techniques and urban geography.

The Institute has no building as such but operates from offices in The Royal Geographical Society at 1 Kensington Gore, London, S.W.7. Enquiries should be addressed to the Administrative Assistant.

Research

Geographical research leading to the award of higher degrees may in theory be undertaken by any graduate in geography. In practice, however, nearly all full-time research involves financial support from public funds, university scholarships or charitable trusts, and the competition for the limited awards available means that a high standard must be achieved in the first-degree examination before an application can be seriously considered. Research assistance in the public sector is largely organized through the Department of Education and Science and the government research councils, two of which will consider applications from geographers.

The Natural Environment Research Council was one of four research councils set up by the government following the recommendations of the Trend Committee on research activities in Great Britain. N.E.R.C. supervises the work of a number of bodies concerned with our environment, such as the Nature Conservancy, the Institute of Geological Sciences and the Institute of Hydrology, and it makes postgraduate awards in the environmental sciences to enable promising young scientists to continue training beyond a first degree. It also makes higher awards to those who have already had a postgraduate training. It has seven training award subcommittees to assist it in its work. Geographers hoping to undertake research in geomorphology, meteorology, climatology, hydrology, biogeography, oceanography, or land use, will therefore make application for assistance to N.E.R.C. Full details of the various research studentships, advanced course studentships and research fellowships offered by N.E.R.C. may be obtained on application to the Secretary, The Natural Environment Research Council,

Alhambra House, 27-33 Charing Cross Road, London, W.C.2.

The Social Science Research Council was set up by the government following the recommendations of the Heyworth Committee on Social Studies. The chief objects of the Council are to encourage and support research, and to make grants to students for postgraduate training, in the social sciences. The Council is assisted in its work by ten subject committees, one of which is concerned with human geography and planning. Geographers wishing to undertake research in economic, political, social, historical or other human aspects of the subject will therefore make application for assistance to S.S.R.C. Full details of the various studentships and fellowships offered by S.S.R.C. may be obtained on application to the Secretary, Social Science Research Council, State House, High Holborn, London W.C.1.

The Department of Education and Science offers studentships for full-time postgraduate study in subjects within the field of the humanities at universities in the United Kingdom. Four types of studentships are offered. These are state studentships of one year's duration, and major state studentships, Hayter studentships and Parry studentships of up to three years' tenure.

Geography is specifically listed in the Hayter and Parry Schemes. The former deals with Oriental, Slavonic, East European and African studies, and the latter Latin-American studies. Candidates for these awards should normally possess or intend to acquire a working knowledge of the relevant language of the area which they intend to study.

There may be some doubt in other proposed geographical research in the field of the humanities whether S.S.R.C. or D.E.S. is the most appropriate body to approach. For *systematic* investigations, applications would normally be considered by S.S.R.C., but for *regional* investigations ('area studies') a candidate should apply either to S.S.R.C. or D.E.S. (or, if appropriate, the Scottish Education Department), according to whether he is taking mainly social science or humanities options in his proposed postgraduate studies.

A candidate may apply to both D.E.S. and S.S.R.C. where his proposed study falls within the scope of S.S.R.C. and the Hayter or Parry fields. If he fails to obtain a Hayter or Parry award, he will then be considered for an S.S.R.C. studentship.

Full details of the D.E.S. postgraduate award scheme may be obtained from the Awards Branch, Department of Education and Science, 13 Cornwall Terrace, Regents Park, London, N.W.1.

University Postgraduate Awards. Most British universities have funds, as result of endowments, gifts and legacies, available for research studentships and fellowships, and these are offered for competition either annually or from time to time. Most awards are restricted or tied to particular subjects, areas, types of applicant, etc., and full details can only be obtained from the registrars of individual colleges and universities. However, a summary statement of the fellowships, scholarships and grants tenable at universities in the United Kingdom is published annually by the Association of Commonwealth Universities at 36 Gordon Square, London W.C.1. An appendix to this publication also contains a short list of awards tenable outside the United Kingdom.

Charitable Trusts. A considerable number of local, regional and national charitable trusts support research activities, and geographers have from time to time benefited by assistance from such sources. No list of particular relevance to geography yet exists but intending research workers in search of financial support would be well advised to check whether any charitable trust exists in the locality of their birth and early education (many trusts are tied to particular areas), or in connection with their parents' occupation (some trusts are tied to particular professions), or with particular reference to a proposed investigation (most trusts are limited in their field of activity). Information can also be sought from the Charity Commission at 14 Ryder Street, St James, London S.W.1. where a register of charities and trusts is maintained.

Index